GOAL PROGRAMMING: METHODOLOGY AND APPLICATIONS

GOAL PROGRAMMING: METHODOLOGY AND APPLICATIONS

by

MARC J. SCHNIEDERJANS
Department of Management,
University of Nebraska-Lincoln, USA

Kluwer Academic Publishers
Boston/Dordrecht/London

Distributors for North America:
Kluwer Academic Publishers
101 Philip Drive
Assinippi Park
Norwell, Massachusetts 02061 USA

Distributors for all other countries:
Kluwer Academic Publishers Group
Distribution Centre
Post Office Box 322
3300 AH Dordrecht, THE NETHERLANDS

Library of Congress Cataloging-in-Publication Data

Schniederjans, Marc J.
Goal programming : methodology and applications / by Marc J. Schniederjans.
p. cm.
Includes bibliographical references and index.
ISBN 0-7923-9558-1 (acid-free)
1. Operations research. 2. Programming (Mathematics). I. Title.
T57.7.S36 1994
358.4'034--dc20 94-49375
CIP

Printed on acid-free paper.

Printed in the United States of America

This book is dedicated to the four individuals
who are most cited in the literature for creating,
nurturing and continuously improving the subject
of *Goal Programming*:

A. Charnes, W. W. Cooper, Sang M. Lee and James P. Ignizio

CONTENTS

LIST OF FIGURES

LIST OF TABLES

PREFACE

If you are a graduate student or researcher who is interested in *goal programming* (GP) research, this book will be of benefit to you. To understand how this book will help you, let me explain the motivation behind its creation.

A few years back I was surprised when I read J. P. Ignizio's 1985 book, *Introduction to Linear Goal Programming* and found not a single reference to the most prolific GP researcher (Sang M. Lee) was cited. For what ever reason relevant GP research is not made available, researchers need a comprehensive source of all GP research publications that they can reference. Without knowing what has been done in GP research, it is impossible to know what is left to do in this very important field of research.

The purpose of the book is to provide the most comprehensive reference guide to GP research that is available to date. This book can be viewed as a very large bibliography but is much more. This book provides the most extensive bibliography of GP books and journal research publications (over a thousand citations) in its appendices that has ever been complied. Recognizing the substantive and maturing nature of the subject of GP, this book's content is devoted to identifying all the available literature on methodology and applications which have been reported during GP's almost forty year history. Having collected GP citations for over twenty years, I have found the volume of existing GP research is presently so substantive that no single book could begin to detail the individual methodologies and applications. As such, this book will focus on overviewing and summarizing the contributions to GP of various research studies, rather than detailing the methodological mechanics and application specifics. The objective of this book is to permit researchers and graduate students to know what has been contributed in the past to GP research so that they will know how best to extend the GP envelop of knowledge into the future. This book also seeks to offer suggestions and ideas to promote further GP research.

This book is not an introductory book on GP and does not contain the basic GP algorithms that are found in such books. Instead, this book seeks to augment basic GP books as a comprehensive guide to finding additional research on those algorithms and applications that are not commonly available in introductory books.

This book is organized into five chapters. Chapter 1 describes GP models and explains their relationships with models other fields of study. This chapter sets a common frame of reference on GP modeling for those who might be not be as familiar with all the conventional forms of the GP models used today. In Chapter 2, subjects on GP modeling strategies are presented. As readers will see in this chapter, issues of conflict are not limited to the GP model alone. Chapter 3 is focused on identifying available GP solution methodologies and Chapter 4 on available applications. These chapters are organized by type of methodology and type of application for quick reference by researchers. Finally, in Chapter 5 a brief discussion is presented on GP past research and trends that exist for future GP research. In each of the these chapters the bulk of the GP journal research studies cited in the literature will be listed for quick reference in the text or in numerous tables throughout this book, with notes to identify the primary purpose of each research study. These tables will help to summarize the content of the articles and serve to allow researchers a means to quickly identify citations that may be of interest to their own research. In Appendix A a listing of over 40 GP textbooks, readings books and monographs is presented. In Appendix B a listing of over 980 GP journal research paper citations are presented. All citations are written in the English language.

The accuracy and completeness of this book is my responsibility. The inclusion criteria for journals and books was simple. If the refereed publication contained the name GP or if it is cited in some other publication as having made a contribution to GP it was eligible to be included. One purposeful omission that I now acknowledge is that paper presentations and nonrefereed publications have, for the most part, been excluded from this book. While the individual citations from readings books have generally not been included in the book, the books themselves that have been cited in the GP literature are cited in Appendix A. The body of knowledge that is contained in this book is refereed and as such has undergone a measure of quality control that nonrefereed presentations and publications do not. The all English requirement (or limitation) is similar to other GP bibliographies. It should be mentioned that many of GP publications that are cited in this book were published in countries other than the US. Despite the fact that the citations in the appendices on which this book represents over 20 years of collecting effort on my part, I acknowledge the possibility that omissions and inaccuracies may exist. For all the errors that this book may contain I apologize for them in advance.

Marc J. Schniederjans

CHAPTER 1. INTRODUCTION TO GOAL PROGRAMMING

This book is designed to provide students, faculty and researchers a perspective and review of the total body of published *goal programming* (GP) research to date. The objective of this book is to present a comprehensive overview of goal programming methodology and applications, past and present, as they are reflected in journal publications and books. It is assumed that readers will have some basic knowledge of GP. With the exception of this chapter, the remaining chapters in this book are fairly independent of each other and can be referenced or read independently of the others.

The purpose of this chapter is to describe the subject called *goal programming* or GP and to distinguish the models, methods and applications as a unique subject of study. Specifically, this chapter seeks to introduce and describe differing types of GP models that will be discussed throughout this book. In addition, the relationship of GP within the fields of *management science/operations research* (MS/OR) and *multiple criteria decision making* (MCDM) will be discussed. This chapter will also introduces the concept of the *life cycle* of GP research as a means of adding to the description of the current state of research on this subject.

DESCRIPTIONS OF GOAL PROGRAMMING MODELS

A Point of Origin

The basic idea of GP has been traced by Romero (1992, p. 2) to a study by Charnes, Cooper and Ferguson (1955) on executive compensation. While the term *goal programming* did not appear in this 1955 article, this paper did present a constrained regression idea that embodies the deviation minimizing approach inherent in GP. According to Romero (1992), it was not until Charnes and Cooper's 1961 linear programming textbook, *Management Models and Industrial Applications of Linear Programming* that the term goal programming appeared. Interestingly, it was not presented as a unique or revolutionary methodology, but as an extension of *linear programming* (LP).

In the Charnes and Cooper (1961, pp. 215-221) book, goal programming was suggested for use in solving unsolvable LP problems (e.g., infeasible LP problems). Indeed, GP was not even cited as a term in the index of the Charnes and Cooper (1961) book.

From this humble beginning as an extension of LP, GP has distinguished itself in the years since as a unique problem solving methodology. Thousands of paper presentations have been given on the subject of GP at annual meetings of professional societies, particularly the *Operations Research Society of America*, the *Institute of Management Science*, the *Institute of Decision Sciences*, and the *International Society of Multiple Criteria Decision Making*. Numerous textbooks, readings books and monographs have been devoted in part or totally to the subject of GP. For a listing of these books see *Appendix A, Textbooks, Readings Books and Monographs on Goal Programming*. Virtually every management science or operations research textbook in the last 20 years has voted a chapter to the subject of GP. Journal research publications on the subject of GP have almost reached the millennia mark. For a listing of these articles see *Appendix B, Journal Research Publications on Goal Programming*.

Goal Programming as an Extension of Linear Programming

Since the origin of GP can be traced to LP, a starting point for the GP model can be found by restating the LP model, its assumptions and modeling notation. One version of the LP model can be stated in what is called the *canonical form*:

$$\text{Minimize: } Z = \sum_{j=1}^{n} c_j x_j$$

$$\text{subject to: } \sum_{j=1}^{n} a_{ij} x_j \geq b_i, \text{ for } i=1, \ldots, m$$

$$x_j \geq 0, \text{ for } j=1, \ldots, n \qquad (1.1)$$

Where the $x_1, x_2, \ldots, x_n$ are nonnegative *decision variables* or unknowns and the $c_1, c_2, \ldots, c_n$ are contribution coefficients that represent the marginal contribution to Z for each unit of their respective decision variable. This LP model seeks a single objective or goal of minimizing the *objective function* or Z

function. In the model the objective function is subject to a set of m *constraints*. In the constraints, the a_{ij}, where $i=1, 2, \ldots, n$ and $j=1, 2, \ldots, m$, are *technological coefficients* that represent the per unit usage by x_j of the *right-hand-side coefficient* of b_i.. In this model the n decision variables are required to be non negative.

The LP model implicitly requires the following assumptions (Fang and Puthenpura 1993, pp. 3-4):

1. *Proportionality assumption*: Each unit of each decision variable x_j contributes c_j units to the objective function and a_{ij} units in the ith constraint.

2. *Additivity assumption*: The contribution to the objective function and the technological coefficients in the constraints are independent of the values of the decision variables.

3. *Divisibility assumption*: Decision variables are permitted to be noninteger or have fractional values.

4. *Certainty assumption*: All parameters, a_{ij}, b_i. and c_j must be known with certainty.

The LP model in (1.1) permits the possibility of positive deviation (called *surplus* in LP terminology) from the right-hand-side coefficients in the model, since the sum of the products in the left-hand-side can be greater-than any b_i. It is also true that LP constraints can be stated as less-than or equal-to expressions. In this case, the LP model can permit negative deviation (called *slack* in LP terminology) from b_i. Any LP model can include greater-than or equal-to, equal-to and less-than or equal-to types of constraints.

Regardless of the types of constraints included in an LP model, the mathematical requirements represented by the constraints must be satisfied in order to have a *feasible solution*. It must also be remembered that optimization of the objective function is secondary to finding a feasible solution set of the x_j that will satisfy all of the constraints in a model. When one or more constraints in an LP model finds itself outside or in conflict with the *area of feasible solutions,* we have an *infeasible solution*.

What Charnes and Cooper (1961) suggested in their textbook is that each constraint that makes up an LP model is a separate function, called a *functional*. These functionals are viewed as individual *objectives* or *goals* to be attained. In effect, the b_i are a set of objectives or goals that we must satisfy in order to have a feasible solution. If we subtract b_i from both sides of an equality constraint, we can express the functional as the absolute value of an LP constraint, or:

$$f_i(x) = \left| \sum_{j=1}^{n} a_{ij} x_j - b_i \right| \text{ for } i=1, \ldots, m \tag{1.2}$$

Charnes and Cooper (1961) referring to these functionals as *goals*, suggested that goal attainment is achieved by minimizing their absolute deviation. In those LP problems (i.e., infeasible solutions) where deviation in functionals are inevitable, the best solution occurs by minimizing the deviation. In this way it is possible to obtain a kind of solution where constraints are in conflict with one another. Charnes and Cooper (1961, p. 215) put it this way:

"Whether goals or attainable or not, an objective may then be stated in which optimization gives a result which comes 'as close as possible' to the indicated goals."

Recognizing that deviation from goals will exist in unsolvable LP problems like an infeasible LP problem, Charnes and Cooper (1961, p.217) illustrated how that deviation could be minimized by placing the variables representing deviation directly in the objective function of the model. This allows multiple (and sometimes conflicting) goals to be expressed in a model that will permit a solution to be found. Multiple and conflicting goals are what many scholars use as a distinguishing characteristic to describe how a GP model differs from an LP model. While Charnes and Cooper did not present a general GP model statement in their 1961 book, a generally accepted statement of this type of GP model was presented in Charnes and Cooper (1977):

$$\text{Minimize: } Z = \sum_{i \in m} (d_i^+ + d_i^-)$$

$$\text{subject to: } \sum_{j=1}^{n} a_{ij} x_j - d_i^+ + d_i^- = b_i, \text{ for } i=1, \ldots, m$$

$$d_i^+, d_i^-, x_j \geq 0, \text{ for } i=1, \ldots, \text{m; for } j=1, \ldots, n \tag{1.3}$$

Where d_i^+ is called a *positive deviation variable* and d_i^- is called a *negative deviation variable*. The substantially useless value of Z is the summation of all deviations. The statement in (1.3) that i is an element of the m possible positive and negative deviation variables is to imply that choice in the selection of deviation variables to be included in the objective function is an option. The deviation variables are related to the functionals algebraically where:

$$d_i^+ = 1/2\ [\,|\ \sum_{j=1}^{n} a_{ij}\,x_j - b_i\ |\ + (\ \sum_{j=1}^{n} a_{ij}\,x_j - b_i\ )\,]$$

$$d_i^- = 1/2\ [\,|\ \sum_{j=1}^{n} a_{ij}\,x_j - b_i\ |\ - (\ \sum_{j=1}^{n} a_{ij}\,x_j - b_i\ )\,] \qquad (1.4)$$

The GP model in (1.3) has an objective function, constraints (called *goal constraints*) and the same nonnegativity restriction on the decision variables (and deviation variables) as the LP model. It should be mentioned that some GP researchers (see Ignizio 1985, pp. 25, 30-1) feel that the term objective function is not an accurate term and the terms *achievement function* or *unachievement function* should used in its place. This book will use the term objective function which, as Ignizio (1985, p. 31) admitted, it is the more traditional terminology and because Charnes and Cooper (as well as most others) still refer to it as an "objective" function (see Charnes, Cooper and Sueyoshi 1988).

The GP model in (1.3) can have LP constraints if desired and requires the same model assumptions as the LP model, with the exception of proportionality for c_j. It is also a requirement that $d_i^+ x d_i^- = 0$ must always hold true. Both the decision variables and the c_j parameters are removed from the objective function. The fact that there are no decision variables in an GP model's objective function is one of the unique characteristics that distinguishes a GP model from other quantitative methods. We will discuss other necessary modeling assumptions, limitations and requirements in Chapter 2.

Extensions of the Goal Programming Model

From the time of the Charnes and Cooper (1961) book to 1966, few scholars published research on GP. The only journal research publications during this time include Chambers and Charnes (1961), Charnes and Cooper (1962),

Charnes, Cooper and Ijiri (1963), Joksh (1964), and Charnes and Stedry (1966). A major development in the extension of GP models did occur during this period with the publication of the 1965 book by Y. Ijiri, *Management Goals and Accounting for Control*. According to Lee (1972, p. 17), Ijiri's (1965) book introduced the *generalized inverse* solution approach. We will be discussing GP solution methodology in Chapter 3. Ijiri also introduced *preemptive priority factors* as way of ranking goals in the objective function of the GP model and established the assigning of *relative weights* to goals in the same priority level. The ideas of weighting or ranking goals are two very different concepts and has fostered two different types of GP models.

Charnes and Cooper (1977) stated the *weighted GP model* as:

$$\text{Minimize: } Z = \sum_{i \in m} (w_i^+ d_i^+ + w_i^- d_i^-)$$

$$\text{subject to: } \sum_{j=1}^{n} a_{ij} x_j - d_i^+ + d_i^- = b_i, \text{ for } i=1, \ldots, m$$

$$d_i^+, d_i^-, x_j \geq 0, \text{ for } i=1, \ldots, m; \text{ for } j=1, \ldots, n \qquad (1.5)$$

Where w_i^+ and w_i^- are nonnegative constants representing the *relative weight* to be assigned to the respective positive and negative deviation variables. The relative weights may be any real number, where the greater the weight the greater the assigned importance to minimize the respective deviation variable to which the relative weight is attached. This model is a nonpreemptive model that seeks to minimize the total weighted deviation from all goals stated in the model. We will discuss methodologies to develop these relative weights in Chapter 2.

While Ijiri (1965) had introduced the idea of combining preemptive priorities and weighting, Charnes and Cooper (1977) suggested the GP model as:

$$\text{Minimize: } Z = \sum_{i \in m} P_i \sum_{k=1}^{n_i} (w_{ik}^+ d_i^+ + w_{ik}^- d_i^-)$$

$$\text{subject to:} \quad \sum_{j=1}^{n} a_{ij}\, x_j - d_i^+ + d_i^- = b_i, \text{ for } i=1, \ldots, m$$

$$d_i^+, d_i^-, x_j \geq 0, \text{ for } i=1, \ldots, \text{m; for } j=1, \ldots, n \qquad (1.6)$$

Where w_{ik}^+, w_{ik}^- ≥ 0 and represent the relative weights to be assigned to each of the $k = 1, \ldots, n_i$ different classes within the ith category to which the non-Archimedean treascendental value of P_i is assigned. The P_i are the preemptive priority factors who serve only as a ranking symbol that can be interpreted to mean that no substitutions across categories of goals will be permitted. It is assumed that the ordering of deviation variables in an objective function, will be minimized in order, where $P_i > P_{i+1} > P_{i+2}$ >>> and so on for as many priorities as may exist in a model. It is also assumed that no combination of relative weighting attached to the deviation variables can produce a substitution across categories in the process of choosing the x_j.

Charnes and Cooper (1977) pointed out the fact that the GP model in (1.6) can allow us to move completely away from weighting deviation variables towards an *absolute priority structure*, where each of the functionals or goals are given a separate priority. Sometimes called the *lexicographic GP model* (see Iserman 1982, Sherali 1982, and Ignizio 1983), the model has no weights, only a preemptive ranking for each of the goals in the model, which can be stated as:

$$\text{Minimize: } Z = \sum_{i \in m} P_i \; (d_i^+ + d_i^-)$$

$$\text{subject to:} \quad \sum_{j=1}^{n} a_{ij}\, x_j - d_i^+ + d_i^- = b_i, \text{ for } i=1, \ldots, m$$

$$d_i^+, d_i^-, x_j \geq 0, \text{ for } i=1, \ldots, \text{m; for } j=1, \ldots, n \qquad (1.7)$$

While several other types of GP models have been suggested, the overwhelming majority of the GP models that appear in the journal literature fit into a form similar to (1.3), (1.5), (1.6) or (1.7). Other variants of GP models such as the *MINMAX GP model* (Romero 1991, p. 5), fall as much under the subject of LP as GP. This book seeks to limit its domain of coverage just GP models that have appeared in journal research publications or GP books.

The *nonpreemptive, nonweighted GP model* in (1.3) is a distinct model that has repeatedly appeared in the literature since the publication of the Charnes and Cooper (1961) book. The *nonpreemptive weighted GP model* in (1.5) and the *preemptive lexicographic GP model* in (1.7) can viewed as the two extreme types of GP models in which virtually all GP modeling are derived. The GP model in (1.6) is simply a combined version of the two extreme types. These GP models, or combinations of them, will be the focus of discussion in this book.

RELATIONSHIP OF GP TO MS/OR

Management science (MS) and *operations research* (OR) can both be defined as a field of study for the application of mathematical analysis to solve managerial problems (Kwak and Schniederjans 1987, p. 1). These fields of study consist of a collection of mathematical algorithms and logic structures that are used to solve problems. OR/MS can trace their origin of industrial application of methodology to the development of linear programming in the 1940's (Cook and Russell 1993, p. 5). As stated in the last section, GP can also trace its point of origin from the development of LP. Just as GP extended LP (by solving unsolvable LP problems, adding weighting, adding priorities, etc.), GP extended itself by reengineering many of the prior single objective LP models with multiple objectives.

There is a direct relationship between the topics in MS/OR and GP. MS/OR journal research and books usually contain the use of one or more mathematical or logic methodologies that exists in the field. Typical MS/OR topics covered in such publications included those listed in Table 1-1. It is assumed that readers are familiar with the MS/OR topics listed in Table 1-1. Most of the MS/OR topics are special case LP methodology or LP models. Logically, then as GP developed, many researchers who had used single objective LP models, extended their model's capabilities to include multiple objectives by reformulating it as a GP model. A early illustration of just how this could be accomplished is presented in Clayton and Moore (1972).

Table 1-1. MS/OR Topics and Their Related GP Topics

MS/OR Topic	Related GP Topic
Linear Programming (LP)	Linear Goal Programming (GP)
LP Duality	GP Duality
LP Sensitivity Analysis	GP Sensitivity Analysis
Integer LP	Integer GP
Nonlinear Programming	Nonlinear GP
Transportation and Assignment Models	GP Transportation and GP Assignment Models
Network Models	GP Network Models
Dynamic Programming Models	GP Dynamic Programming Models
Game Theory Models	GP Game Theory Models
Markov Analysis Models	GP Makov Analysis Models

As can be seen in Table 1-1 there is a related GP topic to each of the MS/OR topics. Since much of the methodologies used to solve LP problems, like the *simplex method*, *duality*, *sensitivity analysis*, *nonlinear programming*, etc., could work on GP problems with minor revisions to the algorithms, most LP methodologies were inevitably converted to a GP equivalent. We will be discussing these GP methods in Chapter 3. Special case LP models, like those for the *transportation method* models, *assignment method* models, *network models*, etc., have LP model equivalents. It is, therefore, an easy step to convert the LP models of these special case problems into GP models by simply adding additional objectives to make the resulting models multi-objective. For example, Drandell (1977) for example, converted one of his own prior LP models used to solve an insurance management problem to a GP model with multiple objectives. We will be discussing these special case formulations in Chapters 2 and 4.

One of the problems scientists have in categorizing methodologies is the fact that one method can be categorized in more than one way. This can lead to confusion and the question as to whether GP is a MS/OR methodology or not. The answer is yes, GP is one of a set of MS/OR mathematical techniques that can be used to solve managerial problems when multiple objectives are present. For this reason, GP is included as a unique problem solving methodology in virtually every MS/OR book on the market today.

RELATIONSHIP OF GP TO MCDM

Multiple criteria decision making (MCDM) is a term used to describe a subfield in operations research and management science. Zionts (1992) generally defined MCDM as a means to solving decision problems that involve multiple (sometimes conflicting) objectives. While that definition also applies to GP, MCDM is a substantially broader body of methodologies of which GP is a small subset.

The various points of origin, methodology and future directions for MCDM can be found in Starr and Zeleny (1977), Hwang, Paidy, and Yoon (1980), Rosenthal (1985), Steuer (1986) and more recently in Dyer, Fishburn, Steuer, Wallenius and Zoints (1992). MCDM's mathematical relationship to GP has been substantially described in a variety of publications including Romero (1991) and Ringuest (1992). Readers are encouraged to review these materials if methodological differences between GP and MCDM are the focus of the reader's research. Consider also reading the entire *Management Science* special issue (Vol. 30, No. 1, 1984) and the entire *Computers and Operations Research* special issue (Vol. 19, No. 7, 1994) both on MCDM, or reviewing any issue of *Multi-Criteria Decision Analysis* journal by John Wiley and Sons.

On a conceptual level the relationship of MCDM and GP can be seen in what Zionts (1992) calls the four subareas that make up MCDM. These four subareas that comprise MCDM are listed in Table 1-2. According to Zionts (1992) the subarea of *multiple criteria mathematical programming* refers to

Table 1-2. MCDM Subareas and Their Related GP Topics

MS/OR Subarea	Related GP Topic
Multiple Criteria Mathematical Programming	Linear Goal Programming
Multiple Criteria Discrete Alternatives	Integer Goal Programming and Zero-One Goal Programming
Multiattribute Utilitiy Theory	Linear Goal Programming, Nonlinear GP and Fuzzy GP
Negotiation Theory	Interactive Goal Programming and GP Game Theory Models

solving primarily deterministic, mathematical programming problems that have multiple objectives. Linear goal programming is one of the many methodologies that are considered a significant contributor to this subarea of MCDM. Indeed, Dyer, Fishburn, Steuer, Wallenius and Zoints (1992) suggest that the development of GP was a beginning point for MCDM, particularly this subarea. How can one distinguish a GP model from the other multiple criteria mathematical programming models? In most cases, the MCDM models in this subarea have decision variables in their objective function, while GP models do not.

Multiple criteria discrete alternatives according to Zoints (1992) is used a problem situation characterized by a matrix where the rows are discrete alternatives in which a choice must be made and the columns constitute objectives. One type of GP methodology we will discuss in Chapter 3 is called *integer goal programming*. Integer GP is a collection of GP algorithms that generate integer or binary (all so called *zero-one GP* or 0-1 GP) solutions for GP problems. Many tabular problems have been formulated as an integer GP problem. For example, Schniederjans and Hoffman (1992) formulated a multinational acquisition problem using a zero-one GP model where the decision choice matrix was made up of selecting multi-national firms to acquire (i.e., the alternatives) based on acquisition criteria (i.e., the objectives).

The subarea referred to as *multiattribute utility theory* from Zoints (1992), involves problems that have probabilistic based outcomes. These outcomes are sometimes expressed as utility functions that are in turn used to weight rankings of alternatives. As we will be discussing in Chapter 3, *nonlinear GP* and *fuzzy GP* methods exist to deal with probabilistic or utility based decision making. In addition, many multiattribute utility theory problems can be converted into deterministic problems and modeled using linear GP. One of the best examples that illustrates the conversion process can be found in Ringuest and Gulledge (1983).

The final subarea of MCDM Zoints (1992) calls *negotiation theory*. The type of problem situation that uses this approach to MCDM involves one or more decision makers who will reach a decision involving multi-criteria information using mechanisms like voting or majority rule. Unique to this type of decision making is the characteristic of where interaction and negotiation takes place by the decision makers to include new are altering information on which to reach a decision. In other words, information changes through negotiation and the new information is shared through interaction among the decision makers. *Interactive GP* is one methodology that can be applied to problem solving where decision makers seek to interact based on new information to reach a final decision. There have also been several *GP game theory* models offered in the literature that fit negotiating theory problems that possess gaming elements (see Hannan 1982b).

Each of these four MCDM subareas really represent a collection of methodologies to be used in a highly varied set of decision environments. While GP modeling can contribute to each subarea, not all MCDM methodology or problem situations can be handled by GP. The presentation here is just to make the point that GP is versatile enough to make some contribution to every subareas that makes up MCDM. In summary, GP is a subset methodology that fits substantially in the multiple criteria mathematical programming subarea, but can provide service to a select group of problems found in the other three subareas.

In conclusion, the relationship of GP to MS/OR and MCDM is one of subordination. As can be seen in Figure 1-1, GP is subordinated within the field of MCDM, which in turn is subordinated within the field of MS/OR. In researching GP literature it is necessary to view GP as a methodology that supports decision making in the various subareas within MCDM. Many of the GP citations in *Appendix B* use the title multi-criteria programming rather than GP. Such labeling can, and probably has lead to researchers overlooking relevant GP contributions. It is also necessary and helpful to use the various topical areas within MS/OR as a means of locating relevant GP models. In Chapter 2 will be discuss the formulations of most MS/OR methodologies within the context of a GP model.

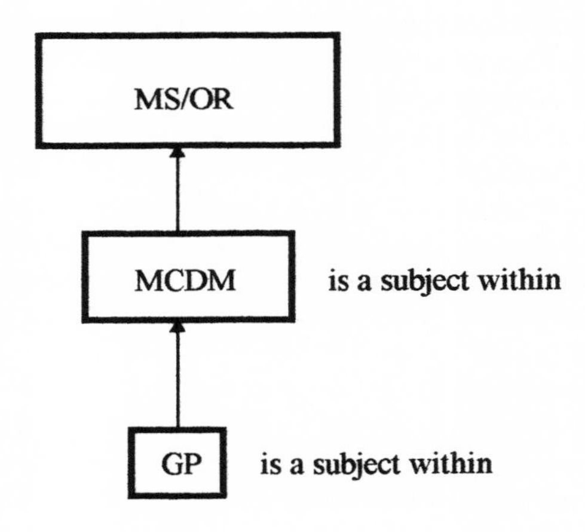

Figure 1-1. Summary Relationship of GP with MS/OR and MCDM

THE LIFE CYCLE OF GP RESEARCH

As previously mentioned, GP's point of origin began with Charnes, Cooper and Ferguson (1955) and now spans 40 years. During that time thousands of GP paper presentations have been made, dozens of papers in reading books have been published and perhaps hundreds of GP papers have been published in non-English journals all over the world. The bibliography of journal publications in *Appendix B* has over 980 citations written in English from journals all over the world. If we plot the frequency of journal publications from 1955 to 1993 (1994 is incomplete as of the writing of this book), their frequency distribution would appear similar to Figure 1-2. The actual numbers of GP journal articles published each year and relevant GP books are listed in Table 1-3.

If we smooth the frequency distribution in Figure 1-2, it appears in the almost exact shape of a classic *life cycle* distribution as pictured in Figure 1-3. Since the distribution being presented consists of GP journal research publications, Figure 1-3 can be viewed as the *life cycle of GP research*. The life cycle of GP research (or any life cycle) goes through four stages of life: Introduction, Growth, Maturity and Decline. The exact positioning of the stages of the life cycle are dictated by the curve's inflection points. Based on these stages, it would appear as though GP research is currently in the Decline stage of its life cycle. If GP research continues in the stage of decline, the subject will cease to exist in a very short period of time. We will revisit the life cycle of GP research in Chapter 5 when we discuss future directions for GP research.

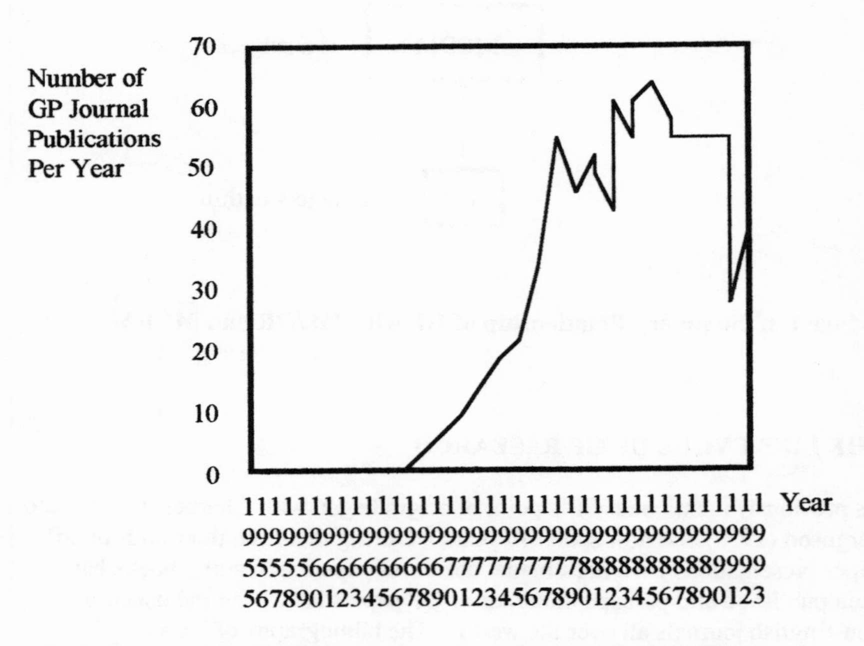

Figure 1-2. Frequency Distribution for GP Journal Publications

Table 1-3. Frequency Listing of GP Journal Publications and Book Titles

Year	No. of Journal Articles	Book Titles with Unique or Significant GP Content*
1955	1	
1956	0	
1957	0	
1958	0	
1959	0	
1960	0	
1961	1	Charnes and Cooper, *Management Models and Ind. Apps. of LP*
1962	1	
1963	1	Graves and Wolfe, *Recent Adv. in Mathematical Programming*
1964	1	
1965	0	Ijiri, *Management Goals and Accounting for Control*
1966	1	
1967	1	Abadie, *Nonlinear Programming*
1968	3	
1969	3	
1970	6	
1971	12	
1972	14	Lee, *Goal Programming for Decision Analysis*
1973	16	Cochrane and Zeleny, *Multiple Criteria Decision Making*
		Zeleny, *Multiple Criteria Decision Making*
1974	19	
1975	22	Zeleny, *Multiple Criteria Decision Making*
1976	29	Ignizio, *Goal Programming and Extensions*
		Keeney and Raiffa, *Decisions with Multiple Objectives*
		Jaaskelainen, *Linear Programming and Budgeting*
1977	32	Starr and Zeleny, *Multi. Crit. Dec. Making-TIMS Studies*
		Van Delft and Nijkamp, *Multi-Crit. Analy. and Reg. Dec. Mkg.*
1978	55	Zoints, *Multiple Criteria Problem Solving*
		Cohon, *Multiobjective Programming and Financial Planning*
1979	43	Lee, *GP Methods for Multiple Objective Integer Programs*
		Fandel and Gal, *Multiple Obj. Dec. Making Theory and Apps.*
		Hwang and Masud, *Multi. Obj. Dec. Making Meth. and Apps.*

*Complete citation for all books can be found in *Appendix A*.

Table 1-3. (Continued)

Year	No. of Journal Articles	Book Titles with Unique or Significant GP Content*
1980	49	
1981	48	Spronk, *Interactive Multiple GP: Apps. to Fin. Planning*
		Morse, *Org. Multiple Agents with Multiple Criteria*
1982	42	Ignizio, *Linear Programming in Single and Multiple Obj. Sys.*
		Lee, *Management by Multiple Objectives*
1983	63	Lee and Van Horn, *Acad. Adm.:Plan., Budgt. and Dec. Making*
		Chankong and Haimes, *Multiobjective Decision Making*
		Hansen, *Essays and Surveys on Multi. Criteria Dec. Making*
		Rekanitis, Ravindran and Ragsdell, *Eng. Opt.: Meth. and Appl.*
1984	56	Schniederjans, *Linear Goal Programming*
		Depontin, Nijkamp and Spronk, *Macro-Econ. Plan. with Conf.*
		Zeleny, *MCDM: Past Decade and Future Trends*
1985	61	Ignizio, *Introduction to Linear Goal Programming*
		Gal and Wolf, *Solving Stochastic LP's via Goal Programming*
		Fandel and Spronk, *Multi. Criteria Decision Meths. and Apps.*
		Haimes and Chankong, *Dec. Making with Multiple Objectives*
		Sawaragi, Nakayama and Tetsuzo, *Theory of Multiobj. Optim.*
1986	66	Steuer, *Multiple Criteria Opt.: Theory, Comp. and Appl.*
1987	57	Lawrence, Guerard and Reeves, *Adv. in Math. Prog. with Fin.*
		Kwak and Schniederjans, *Intro. to Mathematical Programming*
1988	53	Zoints, *Multi. Crit. Math. Prog.: An Update Overview*
1989	50	
1990	51	Lawrence, Guerard and Reeves, *Adv. in Math. Prog. with Fin.*
		Rios-Insua, Se*n. Anal. in Multi-Obj. Decision Making*
1991	55	Romero, *Handbook of Critical Issues in Goal Programming*
		Lieberman, *Multi-Obj. Programming in the USSR*
1992	30	Ringuest, *Multiobjective Optm.: Behavioral and Comp. Cons.*
1993	38	Lawrence, Guerard and Reeves, *Adv. in Math. Prog. with Fin.*
		Sakawa, *Fuzzy Sets and Interactive Multiobj. Optimization*
1994	3**	

*Complete citation for all books can be found in *Appendix A*.

**Incomplete collection for this year.

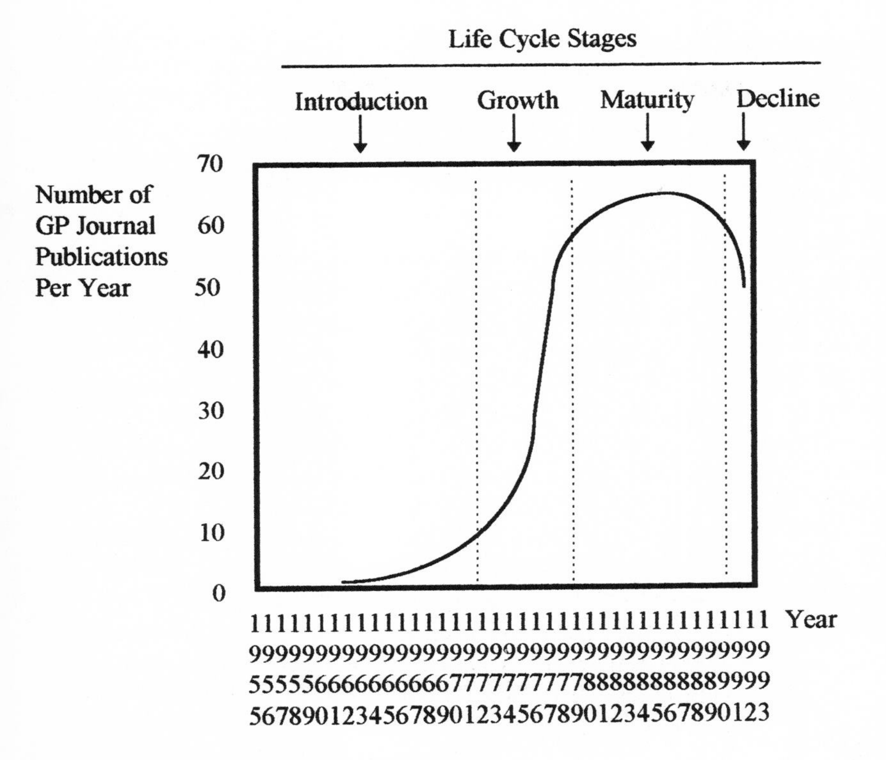

Figure 1- 3. Life Cycle of GP Research

The solution to the problem of declining GP publications is a simple one, that is, to publish more GP research publications. This can only be done by offering journals new and different GP research that makes a significant contribution. To know what is new and different a researcher must know what has been done. To that end this book was created.

SUMMARY

The introductory chapter of any book that deals with an advanced subject usually seems to raise more questions than it answers. This book is no exception. This chapter described several types of GP models (see (1.3), (1.5), (1.6) and (1.7)) that represent the majority of those that appear in GP research. The form and style of modeling notation that will be used throughout this book was introduced in their presentation. This chapter also conceptually described the relationship of GP within the subject areas of MS/OR and MCDM. This book views GP as complementary subset methodology that has application to most multi-objective decision environment subjects discussed in MS/OR and MCDM. The chapter ends with an introduction to the concept of the life cycle of GP research.

It may seem a bit ironic to suggest in an introductory chapter that the life cycle of GP is in decline and may be coming to an end. Others have suggested that GP has already been killed (see Ignizio 1994). The author of this book takes exception with any notion that GP is dead, dying or even has the flu. No body of knowledge or life cycle representing that body of knowledge ever dies. General Dougles MacArthour's put it best when he told the U.S. Congress, "... old soliders never die, they just fade away." GP will never die, but it may fade away if the people who do GP research stop conducting the research. The solution to the problem of declining GP publications is a simple one, that is, to publish more GP research publications. This can only be done by offering journals new and different GP research. To know what is new and different a researcher must know what has been done. To that end this book was created.

To let GP fade away would be a tragic opportunity loss and can be avoided by using some marketing know-how. Product analyst know that a product life cycle can be shifted from a stage of decline to one of growth if something new is added to their products. After almost a thousand journal publications, can there be anything new to add to GP research? You can bet there is! Lets talk about it during the next four chapters.

REFERENCES

All references in this chapter, except those below, can be found in *Appendix B, Journal Research Publications on Goal Programming.*

Charnes, A. and Cooper, W. W., *Management Models and Industrial Applications of Linear Programming*, Vols. 1 & 2, John Wiley and Sons, New York, NY, 1961.

Cook, T. M. and Russell, R. A., *Introduction to Management Science*, 5th ed., Prentice Hall, Englewood Cliffs, NJ, 1993.

Dyer, J. S., Fishburn, P. C., Steuer, R. E., Wallenius, J. and Zoints, S., "Multiple Criteria Decision Making, Multiattribute Utility Theory: The Next Ten Years," *Management Science*, Vol. 38, No. 5 (May 1992), pp. 645-654.

Fang, S. and Puthenpura, S., *Linear Optimization and Extensions: Theory and Algorithms*, Prentice Hall, Englewood Cliffs, NJ, 1993.

Ignizio, J. P., "Who Killed Goal Programming and Why?," paper presentation, 1st International Conference, Multi-Objective Programming and Goal Programming: Theories and Applications, University of Portsmouth, England, June, 1994.

Ijiri, Y., *Management Goals and Accounting for Control*, Rand Mc Nally, Chicago, IL, 1965.

Kwak, N. K. and Schniederjans, M. J., *Introduction to Mathematical Programming*, Krieger Publishing, Malabar, FL, 1987.

Rosenthal, R. E., "Principles of Multiobjective Optimization," *Decision Sciences*, Vol. 16, No. 2 (1985), pp. 133-152.

Starr, M. K. and Zeleny, M., "MCDM-State and Future of the Arts," in Starr, M. K. and Zeleny, M., eds., *Multiple Criteria Decision Making: Studies in the Management Sciences*, North-Holland Publishing, Amsterdam-New York, 1977, pp. 5-29.

Steuer, R. E., *Multiple Criteria Optimization: Theory, Computation, and Application*, John Wiley and Sons, New York, NY, 1986.

Zoints, S., "Some Thoughts on Research in Multiple Criteria Decision Making," *Computers and Operations Research*, Vol. 19, No. 7 (1992), pp. 567-570.

CHAPTER 2. GOAL PROGRAMMING MODEL FORMULATION STRATEGIES

Controversy is a part of any modeling effort, particularly goal programming (GP) modeling. Unfortunately controversy in the way GP models are formulated and presented in the literature has undoubtedly lead to many useful and potentially great models being rejected in the review process for publication. There is need for both the creators of GP models and journal reviewers to understand some of the basics and confusing issues of GP model formulation that exist in the literature.

The purpose of this chapter is to review issues related to GP model formulation strategies that have appeared in the literature. This review of issues and survey of available literature may help researchers to avoid common pitfalls in the practice of GP model formulation. In addition this chapter seeks describe a synergistic approach for GP with other management science/operations research models as a strategy for publication. This chapter also reviews the literature on a number of model structural element methodologies as tactics for improving GP models.

A GOAL PROGRAMMING MODEL FORMULATION PROCEDURE

To formulate any of the GP models presented in Chapter 1 will include some, if not all of the following steps that can be found in any basic management science/operations research (MS/OR) or GP book (see Schniederjans 1984, pp.71-72): (1) define the decision variables, (2) state the constraints, (3) determine the preemptive priorities if need be, (4) determine the relative weights if need be, (5) state the objective function, and (6) state the nonnegativitiy or given requirements. Regardless of whose GP book or article is used, these basic steps should be a part of any GP model formulation strategy. Unfortunately at each step in the formulation process it is possible to misformulate a model. The issues that are often overlooked and give rise to a misformulation are the subject of the next section of this chapter.

ISSUES RELATED TO GOAL PROGRAMMING MODEL FORMULATION

GP model formulations can be in error. Unfortunately, what one researcher may call poor modeling (i.e., modeling error) may be viewed by some other researcher as good modeling. Indeed, anyone can find error in someone else's GP model if they want to. Since those who create GP models are generally the same who review those models, care must be exercised in judging the poorness or goodness of any GP model.

Numerous GP publications have dealt with difficulties and criticisms on the use of GP models. Some of the more notable comments are in Harrald, Leotta, Wallace and Wendell (1978), Hannan (1980), Zeleny (1981), Alvord (1983), Rosenthal (1983), Hannan (1985), Ignizio (1985), Gass (1987), Romero (1991), and Min and Storbeck (1991). From these studies and many others, several modeling issues seem to generate the most controversy. These issues include *dominance* in GP solutions, *inferiority* in GP solutions, *efficiency* in GP solutions, *naive relative weighting* in GP models, *incommesurability*, *naive prioritization* in GP models, *redundancy* and others.

Dominance, Inferiority and Efficiency in GP Solutions

Just as in linear programming (LP), GP models can have multiple solutions. Unlike LP, the GP model can permit a variety of alternative solutions that may allow at least one of the model's goals to be improved without worsening or degrading the others. Zeleny (1981) and other GP researchers feel such a situation represents a major defect in GP modeling.

A *dominated* solution occurs in a GP, if and only if, an alternative feasible solution can be found that will not reduce deviation in an objective function while reducing deviation of some other goal. Cohon (1978, p.70) referred to a *nondominated* solution as one where no other feasible solution existed that would improve one goal without degrading other conflicting goals. A dominated solution can be called an *inferior solution* because other superior solutions that yield better satisfying answers exist. Likewise a nondominated solution can be called a *noninferior solution* because it represents the best solution and not one that is inferior to any other.

This issue of *efficiency* in GP solutions is closely related to dominance and inferiority. According to the Italian economist Vilfredo Pareto (Pareto 1971), efficiency is at an optimal level if the economic situation of a group of people can not be improved without worsening or degrading the economic situation of any one person who makes up the group. This type of optimality is called *Paretian efficiency* (Romero 1991, p. 13). A GP solution is said to be paretian efficient if no other feasible solution can achieve the same or better solution for the group of goals that exist in the objective function, while also being better for one or more other individual objectives that exist in the model. So a GP *efficient solution* must be nondominated solution and a noninferior solution. A GP *inefficient solution* is a dominated solution and a noninferior solution.

One issue in GP model formulation that is directly related to the issues of dominance, inferiority and efficiency concerns the use of LP constraints in GP models to restrict decision variable values. It is certainly a fact that GP permits the inclusion of LP constraints into GP models. Moore, Taylor, Clayton and Lee (1978) referred to the LP constraints included in GP models as *system constraints*. Ignizio (1985, p. 23) calls these constraints a set of *rigid constraints*. Regardless of what they are called they can be included in the formulation of any GP model.

The use of LP constraints may have contributed to confusing GP modeling issues. What appears to be a misrepresentation is how dominance, inferiority and efficiency is represented in the literature. The few tangible examples that appear in the literature seem to require LP constraints that restrict the decision variables in a GP model, in order for dominance, inferiority and inefficiency to be present in a GP solution. For example, Romero (1991, p.15) presented the following GP model formulation:

Minimize: $Z = d_1^- + d_2^- + d_3^-$

subject to:
$$x_1 - d_1^+ + d_1^- = 5$$
$$x_2 - d_2^+ + d_2^- = 2$$
$$0.9x_1 + x_2 - d_3^+ + d_3^- = 8$$
$$x_1 \leq 6$$
$$x_2 \leq 5$$
$$x_1 + x_2 \leq 9$$
$$d_i^+, d_i^-, x_j \geq 0, \text{ for } i=1, \ldots, 3; \text{ for } j=1, \ldots, 2 \qquad (2.1)$$

The graphic GP solution for this model is presented in Figure 2-1 and numerically presented under the 2.1 Solution column in Table 1-1. In Figure 2-1 the best solution occurs at point *a* on the graph. According to Romero (1991, p. 16) that solution is dominated, inferior and inefficient by all the other points in the shaded region of *a*, *b* and *c*. This dominance can be proved with a test. One test for dominance (which reveals inefficiency and inferiority) in GP solutions by Masud and Hwang (1981) involves maximizing the positive deviation variables. In this case the deviation variables are given a negative weight, which by minimizing a negative weight achieves a maximization of positive deviation. Applying this method to (2.1) we have a revised preemptive model:

Minimize: $Z = P_1 \, (d_1^- + d_2^- + d_3^-) + P_2 \, [(-d_1^+) + (-d_2^+) + (-d_3^+)]$

subject to:

$$x_1 - d_1^+ + d_1^- = 5$$
$$x_2 - d_2^+ + d_2^- = 2$$
$$0.9x_1 + x_2 - d_3^+ + d_3^- = 8$$
$$x_1 \leq 6$$
$$x_2 \leq 5$$
$$x_1 + x_2 \leq 9$$
$$d_i^+, d_i^-, x_j \geq 0, \text{ for } i=1, \ldots, 3; \text{ for } j=1, \ldots, 2 \qquad (2.2)$$

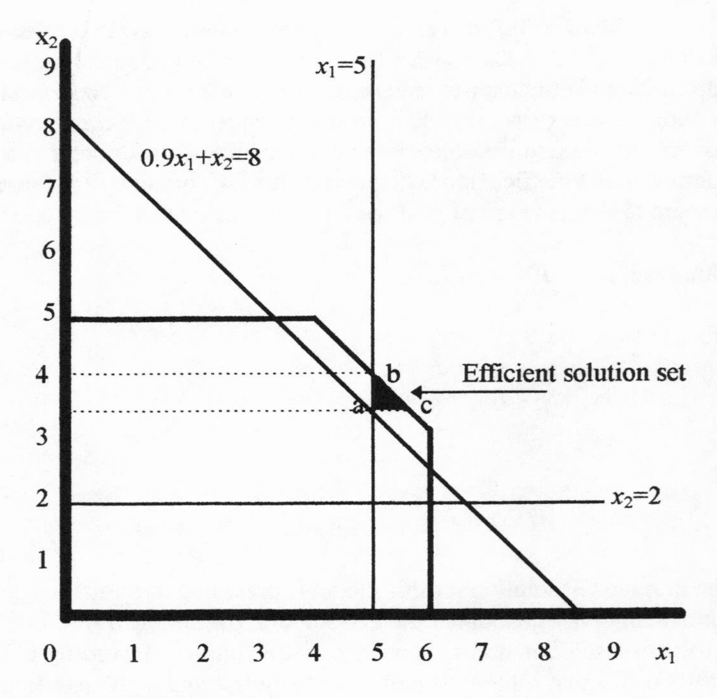

Figure 2-1. Set of GP Efficient Solutions

Table 2-1. Solutions for a Dominated GP Problem

Unknowns	(2.1) Solution	(2.2) Solution	(2.3) Solution*
Z	1.5	2.5	8.4
x_1	5.0	5.0	6.0
x_2	3.5	4.0	5.0
d_1^-	0.0	0.0	0.0
d_2^-	0.0	0.0	0.0
d_3^-	0.0	0.0	0.0
d_1^+	0.0	0.0	1.0
d_2^+	1.5	2.0	3.0
d_3^+	0.0	0.5	2.4

*Other remaining deviation variables not presented.

The solution for this model is also presented in Table 1-1 under the (2.2) Solution column. Since the values for the negative deviation variable are still at zero (an optimal solution for the goals in this model), the (2.2) Solution has not degraded the original problem's goals. Also, we have more positive deviation in variables (i.e., 2>1.5 and 0.5>0.0 for the d_2^+ and d_3^+ , respectively) in the (2.2) Solution, and since we seek to maximize deviation in model (2.2) it represents a solution that is better or dominates the (2.1) Solution.

The problem with this argument is that the existence of the LP constraints in this type of GP model is chiefly the cause of the dominated solution. By converting the LP constraints into goal constraints using Table 2-2 as a guide (from Ignizio 1985, pp. 23-24), this problem will not generated a dominated solution. The GP model for this problem is presented in (2.3) and its solution in Table 2-1 in the (2.3) Solution column. If we accepted the notion that more is better as in the (2.2) Solution, then even more is better in the (2.3) Solution. As we an see, the desire to maximize the positive deviation variables has a higher achievement level, while not degrading the negative deviation variables from zero. More over, when the Masud and Hwang (1981) dominance test is applied, the solution for the model in (2.3) is nondominated, efficient and noninferior. Given that the conversion of the LP constraints to goal constraints is correct, how can this outcome be true? It can be true in GP models where all the constraints are goal constraints and that they are included in the objective function. It is also consistent with the observations by Cohon (1978), that GP models that possess some positive deviation, while minimizing negative deviation variables helps to guarantee nondominated solutions.

Table 2-2. Conversion of LP Constraints to Goal Constraints

LP Constraint	Goal Constraint equivalent	Deviation Variable to be Minimized in the Objective Function
$f(x_j) \geq b_i$	$f(x_j) - d_i^+ + d_i^- = b_i$	d_i^-
$f(x_j) \leq b_i$	$f(x_j) - d_i^+ + d_i^- = b_i$	d_i^+
$f(x_j) = b_i$	$f(x_j) - d_i^+ + d_i^- = b_i$	$d_i^+ + d_i^-$

Minimize: $Z = P_1\ (d_1^- + d_2^- + d_3^-) + P_2\ (d_4^- + d_5^- + d_6^-)$

subject to:

$$\begin{aligned} x_1 - d_1^+ + d_1^- &= 5 \\ x_2 - d_2^+ + d_2^- &= 2 \\ 0.9x_1 + x_2 - d_3^+ + d_3^- &= 8 \\ x_1 - d_4^+ + d_4^- &= 6 \\ x_2 - d_5^+ + d_5^- &= 5 \\ x_1 + x_2 - d_6^+ + d_6^- &= 9 \\ d_i^+, d_i^-, x_j \geq 0, \text{ for } i=1, \ldots, 6&; \text{ for } j=1, \ldots, 2 \qquad (2.3) \end{aligned}$$

In summary, it might be advisable to avoid issues of dominance, inferiority and inefficiency by converting LP constraints to goal constraints. This can be accomplished by simply reformulating LP constraints as goal constraints and placing them at the top priority of the model (Ignizio 1985, pp. 23-24). This formulation strategy does not guaranty to prevent all of the problem causing issues discussed here from occurring, only to reduce their occurrence in GP modeling. As such, it might also be advisable to run one or more tests for dominance or efficiency to confirm such problems are not present in the model formulation.

The debate on the seriousness of the presents of dominance, inferiority and inefficiency in GP solutions has lead many GP researcher to question the worthwhileness of GP as a multicriteria method (Romero 1991, p.14). For the single most comprehensive discussion of the prior literature on both the pro's and con's of these issues in GP modeling, see Min and Storbeck (1991). A general listing of citations on these subjects can also be found in Table 2-3.

Table 2-3. GP Citations on Dominance, Inferiority and Inefficiency

Reference	Notes and Comments on What Reference Provides
Cohon (1978)	a variety of perspectives on dominance
Cohon and Marks (1975)	a classic statement on dominance
Hannan (1985)	a basic review of all three issues
Hannan (1980)	a test for efficiency
Ignizio (1981c)	a discussion on GP efficiency
Kornbluth (1973)	a classic argument on efficiency
Masud and Hwang (1981)	a test for inferiority
Min and Storbeck (1991)	pro's and con's on all three issues
Romero (1991, Chapter 2)	illustrative examples and discussion on all three issues with several tests for efficiency and dominance
Romero and Rehman (1983)	a test for efficiency
Zeleny (1981)	states inherent inferiority of GP models

Naive Relative Weighting, Incommesurability, Naive Prioritization, and Redundancy in GP Model Formulations

The four GP model formulation issues of *naive relative weighting*, *incommensurability*, *naive prioritization*, and *redundancy* are often related to one another. In fact, any one of these issues can cause formulation problems that in turn often causes the other issues to brought up as criticisms against GP models. Its even possible for a GP model can actually suffer from all four of these problems at one time.

In any of the weighted GP models in Chapter 1 (see (1.5) and (1.6)) the *relative weights* establish the importance of goals to which they are attached. If the weights do not accurately reflect the true proportioned weight that rightfully exists in the decision environment that is being modeled, then we have a situation of *naive relative weighting*. Since relative weights are often viewed as a type of utility function, Rosenthal (1983) has argued that weights will almost never reflect the true economic environment they are trying to describe.

The negative impact of naive relative weighting can be minimized by putting more effort into their calculation. The use of such weighting methods as the *analytic hierarchy process* (Satty 1980), *conjoint analysis* (Green and Srinivasan 1990), and even multiple regression analysis can help to improve the accuracy of weighting to reflect the true decision environment a model seeks to describe. We will discuss these and other methods further, in a later section of this chapter.

A related issue to relative weighting, though not in a cause-and-effect relationship, is that of *incommensurability* of goal constraints. In a weighted GP model goal constraints are often used to model very different types of goals. Minimizing a budgetary goal constraint whose a_{ij} and b_i are measured in terms of dollars is incommensurable or not in the same measure of units as a human resource goal constraint whose a_{ij} and b_i are measured in terms of humans to hire for job. When both are included in a weighted GP model its like mixing apples and oranges together. Mathematically, the relative magnitudes of measures taken from differing populations (i.e., dollars vs. humans) bias the solution process in favor of the parameters what will yield the largest reduction in deviation. While the relative weights can sometimes used to adjust the model for parameter magnitudes, it is very difficult to validate the nonbias in such GP models. Such bias can also completely render any relative weighting of goals irrelevant since the weighting is used in combination with the parameters to determine the inclusion in the solution process. This point has been well illustrated in the literature (see Romero and Rehman 1984a).

The impact of incommensurability issues in GP modeling can be minimized in a variety of ways. De Kluyver (1979), Zeleny (1982, pp. 315-333) and Romero (1991, pp. 35-43) have all suggested scaling methods that normalize the goal constraints in a GP model. This normalization has the effect of equating the magnitudes of constraint parameters with little or no violation of the proportionality assumption in the model, so the issue of incommensurability is eliminated between goal constraints. We will discuss this method later in this chapter. Another way to over come incommensurability is to simply prioritize the individual goals if the decision environment permits such rankings. When a priority system is used, differing measurement mixtures of constraints can be keep separate. That is to say, the apple constraints can be kept at one priority and the orange constraints can be kept separate at another priority. As long as the groupings of apple or orange constraints at each priority are weighted in a way not to again cause incommensurability, this method will solve this problem. The lexicographic GP model (Chapter 1, (1.7)) can not suffer from this incommensurability. Still another way of defending a GP model on this issue is to rationalize the logic of setting weights in the applied context of the decision environment. See the Romero (1985b) note and the justifying reply by Sutchliffe, Board and Cheshire (1985).

In summary, the noweighted, nonpriority GP model of (1.3), the weighted, nonpriority GP model of (1.5), and the weighted, preemptive GP model of (1.6) tend to suffer from the issue of incommesurability. The nonweighted, preemptive priority or lexicographic GP model (1.7) does not suffer from incommesurability since each of its goals are separated at a different priority level. Unfortunately, the preemptive priorities that protect GP models from incommesurability, can themselves bring modeling difficulties to the formulation.

In the preemptive GP models in Chapter 1 (see (1.6) and (1.7)), the prioritization or ranking of goals in a GP model must accurately reflect the decision environment. Failure to establish a true prioritization of goals in a GP model brings with it a criticism of *naive prioritization* (Romero 1991, Chapter 4). A related issue of naive prioritization is that of *redundancy* in GP models (Amador and Romero 1989). *Redundancy* refers to goal constraints that are not considered or redundant in a given solution for a GP model. For example, in the (2.4) GP model, the resulting solution is where $x_1 = 1$, $x_2 = 1$, $P_1 = 0$,

P_2 =99, P_3 =199, P_4 =0.95, and P_5 =0.92. Only the P_1 goal is fully achieved. The fact is the solution was actually determined at the P_1 goal. This resulted in the other goals being *redundant* or not being considered in the solution. A reviewer might suppose a naive prioritization of goals has caused the redundancy of the other goals. This may or may not be true. If the priorities reflect the actual decision process, then the true decision is determined at the P_1 goal level. The fact that other goals are a part of a GP model and that they are not considered in a solution, should not be considered a poor modeling practice. If this were true, every LP model that has a slack or surplus variable (which includes about every model ever formulated) would be considered misformulated. Similar notions of the inappropriateness of setting a number of priorities in a GP model (see Ignizio 1985, p. 32 who suggests a maximum of 5 or 6 priorities in any GP model) are a questionable suggestion. Every modeling situation is different and will require a differing amount of priorities to adequately describe the decision environment in that situation. It should be remembered that the trade-off information revealed by priorities that are not fully satisfied is often more valuable then the solution values for the decision variables. Without some deviation in the lower priority levels trade-off information is not available, and that deviation would include redundant goal constraints.

Minimize: $Z = P_1\,(d_1^+ + d_2^+) + P_2\,(d_3^-) + P_3\,(d_4^-) + P_4\,(d_5^+) + P_5\,(d_6^+)$

subject to:

$$
\begin{aligned}
x_1 - d_1^+ + d_1^- &= 1 \\
x_2 - d_2^+ + d_2^- &= 1 \\
x_1 - d_3^+ + d_3^- &= 100 \\
x_2 - d_4^+ + d_4^- &= 200 \\
x_1 - d_5^+ + d_5^- &= 0.05 \\
x_2 - d_6^+ + d_6^- &= 0.08 \\
d_i^+, d_i^-, x_j &\geq 0, \text{ for } i=1, \ldots, 6; \text{ for } j=1, \ldots, 2 \qquad (2.4)
\end{aligned}
$$

One reason that does not appear to have been given must discussion time in the literature for all the modeling formulation issues is that of *modeling ethics*. Should a GP researcher accurately model a real world application with all its naiveté or should they exercise some mathematical correctness (i.e., political correctness in modeling) just to get the model through a review process for publication? Many professionals in industry are naive of current state-of-the-art mathematical conventions. Real world situations often reflect a more naive approach to modeling. Researchers seeking to model a real world application can find that decision makers don't always include relevant information, weight the importance of information and/or rank information in a way that the mathematical purist might want them to. Is it wrong then, in structuring a model of the real world, to include naive weighting or naive rankings if that is the way it is actually done? Is it wrong to reflect the real world the way it is observed by the researcher or the way in which a client wants it to be modeled? Should a researcher willfully remove real world naiveté from models in an effort to help them get published? These are ethics questions that experienced GP researchers have to be able to answer for themselves. Most researchers will not change the real world naiveté but instead try to change the client or decision making situation to reflect a less naive situation through education. If the researcher can not change the naive situation, then is advisable to mention the possibility of the GP model possessing some naiveté. It is hoped that reviewers of the same GP papers will not be so quick to reject a real world application simply because it accurately portrays a bit too much naiveté in its formulation.

Citations on relative weighting, prioritization and incommensurability can be found throughout the entire *life cycle* of GP literature. Some of the citations focus on methodology or systems to derive weightings or priorities, while other just provide conceptual material or comments on these issues. Table 2-4 presents a listing of some citations that are repeatedly referenced in the literature.

Table 2-4. GP Citations on Relative Weighting, Prioritization and Incommensurability

Reference	What Reference Provides
Choo and Wesley (1985),Cook and Kress (1988), Gass (1986, 1987), Hotvedt, Leushner and Buhyoff (1982), Knoll and Engelberg (1978), Levary and Avery (1984), Norse and Clark (1975), Ng (1987), Phillips (1987), Steuer (1979)	all provide weighting systems
Romero (1985b), Schenkerman (1991), Sherali (1982), Sutchiffe, Board and Cheshire (1985)	all provide comments on weighting systems

Table 2-4. (Continued)

Reference	What Reference Provides
Gass (1986), Lee and Shim (1986), Rubin and Narasimhan (1984), Tiwari, Dharmar and Rao (1986)	all provide priority systems
Kluyver (1979), De Kluyer (1979b), Fishburn (1974), Hannan and Narasimhan (1981), Hannan (1981d), Rae (1974)	all provide comments on priority systems
Harrald, Leotta, Wallace and Wendell (1978)	discusses limitations of GP on commensurability
Ginguest and Gulledge (1983)	discusses prioritization as a means of dealing with incommensurable goals
Cohon and Marks (1975)	argument that GP is limited to situations where goals or target values must be clearly defined
Min and Storbeck (1991)	provides pro's and con's on all three issues
Romero (1991, p. 35-43)	provides comments and teachable examples of all three issues

Other GP Model Formulation Issues

The number of possible GP model formulation issues far exceeds the content capacity of a single book to be able to do them justice. On the other hand, some mention of these other issues will be briefly offered here to serve as a directory for researchers who need additional information on these sometimes publication blocking issues.

Inappropriateness of predetermined goals or targets (Min and Storbeck 1991): Setting *a priori* right-hand-side b_i values (the model goals or targets) in GP models is viewed by some as being too arbitrary. It is felt that this arbitrary setting may lead to problems like dominance or just limits the information that the GP model can provide (Zeleny 1981). To overcome objections it might be advisable to included a sensitivity analysis of b_i Steuer (1976, 1979) that eliminates the uncertainty in the parameter. Another method (Romero 1985a) involves setting impossibly high target values to force nondominated solutions to occur. Still other methods to resolve this issue include those in Narasimhan (1980), Werczberger (1981), and Rakes and Franz (1985).

Failure of GP to identify unbounded solutions (Min and Stoebeck 1991): As Lee (1972, p.122) points out, unbounded solutions can occur in GP models if they are misformulated. Defending points of view can be found in Ijiri's (1965) book, Iginizo's (1976) book, Schniederjans's (1984) book and repeatedly by others in the literature including Hannan (1980, 1985), and Markowski and Ignizio (1983) and Ignizio (1983). This is not a justifiable reason to reject a GP model formulation.

Equivalence of GP and LP models (Romero 1991, pp. 26-29): As Ignizio (1983) points out, GP models are not LP models. Yet it is the LP base of problem situations that has so fueled the *GP life cycle of publications*. GP models can solve LP model problems for exactly the same solution, but can do much more and in different ways. Romero (1991, p. 29) claims that the equivalence in solutions between an LP model and a GP model is not a desirable situation. This has not been the case in the hundreds of GP models that fill the literature with LP model formulations, expressed with multiple objectives using a GP model. It is quite logical to start with a problem formulation as an LP model, recognize LP's limitations to deal with multiple objectives in the decision environment, and then revise the model in terms of GP. Even in the situation where the solution generated by the GP and the LP model are the same, GP methodology can give trade-off information (such as in the case of preemptive priorities) that does not exist in LP models. This is not really a problem that should be used to reject a GP model formulation.

Logical structure of a goal constraint (Romero 1991, pp. 31-33): Romero (1991, pp. 31-32) and others believe that both deviation variables must be included in goal constraints for a correct formulation. This may be true in some situations, but not all. Most experienced consultants know that in the real world there are boundaries that permit absolutely no deviation. To reflect this extremely hard boundary in a decision environment a single deviation variable constraint can be used. Charnes and Cooper's (1961, p. 217) original GP model includes single deviation variable goal constraints. For those unfamiliar with their formulation Schniederjans (1984, p. 70) illustrates the procedure by which single deviation variable constraints are formulated. Unless leaving the one of the deviation variables out of a formulation leads to some other formulation issue, this omission should not be used to reject a GP model formulation.

Formulation of a model that provides an optimal or satisficing solution: Since the GP solution methods employ the same basic optimization methods used in LP solutions (i.e., *simplex algorithms*), their is a component of optimization at work in any GP model (Ignizio 1982, p. 402). It should not viewed as an incorrect statement that GPmodels seek to optimize solutions or decision making from those solutions. Most GP researchers prefer to consider GP solutions as *satisficing*. The concept of satisficing implies that GP seeks a solution that fully satisfies as many goals as possible rather than just optimizing a single goal as in an LP model (Lee 1972, Chapter 1, Schniederjans 1984, p. 68, Romero 1991, p. 14). Based on the *Simonian philosophy* of satisficing (Simon 1955, 1957), GP models are able to solve complex problems with multiple goals in a way that best satisfies those goals, however they are structured. In summary, the terms optimal, satisficing, best, or even approximately rational (Lee 1972, p. 6) can all be used to describe a GP solution and should be viewed as an acceptable description of what the GP model does.

In concluding this section the concluding remarks by Charnes and Cooper (1961, p. 222) seem most appropriate. At the outset of the creation of GP, Charnes and Cooper where careful to warn users in the formulation of GP models of two main points: "(1) the need for considering all factors, including equivalencies, in setting goals and objectives, and (2) the wisdom of doing this as an integral part of the analysis rather than attempting to think all aspects of the problem through without the aid of explicit models." Care in GP modeling and care in the review of GP models should be exercised by all.

FORMULATION STRATEGIES AND TACTICS FOR IMPROVING GP MODELING

A general strategy for the creation or development of new GP models has often began with a methodology other than GP. In this section we will examine the use of MS/OR models as a point of creation for GP modeling. Once a GP model has been formulated, there is also the need to improve its structure as a means of adding value or literary significance to the model. Tactical methodology that can be used to improve model parameters and structure will also be discussed.

MS/OR Model Formulation Strategies

The synergistic impact of combining two or more MS/OR models to solve a complex decision situation is common place in the literature. LP has been shown to be useful in modeling most of the MS/OR models, including *transportation models*, *assignment models*, *network models*, *dynamic programming models*, *game theory models*, and *markov analysis models* (Kwak and Schniederjans 1987, Fang and Puthenpura 1993). GP is able to model LP problems with multiple objectives. Therefore it is only logical that researchers have chosen to use this connection to structure virtually all of the MS/OR models related to LP, as GP models. In addition, as other quantitative methodologies have found their way into MS/OR books, such as *simulation models*, *heuristic models* and *statistical models* (Hillier and Liegerman 1986), these methodologies were also joined with GP models to provide unique information or analytic power to deal with complex problems. In Table 2-5, available GP research publications are listed by their respective MS/OR topic.

There is still considerable GP modeling that can be done on any of these topics. On each of the MS/OR topic areas in Table 2-5 there are hundreds of articles describing these models that date back almost 50 years. The handfuls of GP models referenced in Table 2-5 representing the multiple goal conversion of the MS/OR topic can not possibly have done justice to still unused potential of each of these topics. For example, the topic of *heuristics* is currently being reinvented and applied in a growing number of areas including *artificial intelligence* and *expert systems*(Ignizio 1991, pp. 29-43). It can be assumed, as such, that the topic of heuristics is in an early stage of its own *life cycle* of research. By combining GP models with heuristic models, the effect will be to move the GP life cycle back in time toward a *maturity* or even *growth stage* in its own life cycle. Combining GP with other MS/OR models becomes not only a strategy a modeling ideas for the researcher, but also a strategy of life cycle renewal for the subject of GP.

Table 2-5. MS/OR Topics and Their Related GP Topics

MS/OR Topic	Related GP Model Reference
Transportation Models	Acharya, Nayak and Mohanty (1987), Bit, Biswol and Alam (1993), Kwak and Schniederjans (1979, 1985d), Lee and Moore (1973), Nayak, Basu and Tripathy (1989), Singh and Kishore (1991), Stewart and Ittman (1979)
Assignment Models	Bailey, Boe and Schnack (1974), Charnes, Cooper, Niehaus and Stedry (1969), Lee and Schniederjans (1983), McClure and Wells (1987b), Mehta and Rifai (1976, 1979), Phillips (1987), Schniederjans and Kim (1987), Zanakis (1983)
Network Models	DePorter and Kimberly (1990), Dieperink and Nijkamp (1987), Ignizio (1983b), Ignizio and Daniels (1983), Jandy and Tanczos (1987), Pentzaropoulos and Giokas (1993), Premachandra (1993), Price and Gravel (1984), Price (1978), Qassim and Silveira (1988), Van Hulle (1991a, 1991b)
Dynamic Programming Models	Acharya, Nayak and Mothanty (1987), Basu, Pla and Ghosh (1991), Charnes, Duffuaa and Al-Saffar (1989), Kornbluth (1986)
Game Theory Models	Charnes, Duffuaa and Intriligator (1984), Cook (1976), Hannan (1982b), Ratick (1983), Schniederjans and Kim (1987)
Markov Analysis Models	Zanakis and Maret (1981a, 1981b)

Table 2-5. (Continued)

MS/OR Topic	Related GP Model Reference
Simulation Models	Ashton (1985, 1986), Clayton, Weber and Taylor (1982), Dobbins and Mapp (1982), Lin (1978), Rees, Clayton and Taylor (1985)
Heuristic Models	Chen and Askin (1990), Gabbani and Magazine (1986), Loucks and Jacobs (1991), Mendoza (1986)
Statistical Models	Bajgier and Hill (1984), Chakraborty (1991b), Hattenschwiler (1988), Charnes and Cooper (1975), Charnes, Cooper and Sueyoshi (1986, 1988), Miyajima and Masato (1986), Sueyoshi (1989)

Tactics for Improving GP Model Formulations

Capturing the modeling complexity in a decision environment with multiple objectives can be difficult even for GP models. GP model formulations are often criticized for a variety of reasons including the logic or mathematical rationale of parameters that are used to make up the model. The relative weights, the number of priorities, the actual values for the technological coefficients of a_{ij} and right-hand-side values of b_i all can contain flaws, where only a single flaw can ruin an entire model formulation.

To help minimize these potential flaws a number of methodologies have been suggested in the literature that are useful in improving GP model formulations. Some of these methods include *scaling* or *normalizing* of goal constraint parameters, the use of *analytic hierarchy process* for determining relative weights or priorities, the use of *conjoint analysis* for determining relative weights or priorities, the use of *regression analysis* for determining relative weighting or goal constraint parameter estimation, the use of *logarithmic transformations* of goals and the use of *input-output analysis* for technological parameter estimation.

Scaling or normalizing goal constraint parameters: The suggestions by De Kluyver (1979), Widhelm (1981), Zeleny (1982, pp. 315-333), Gass (1986, 1987) and Romero (1991, pp. 35-43) on scaling methods that normalize goal constraints in a GP model can be used to reduce the bias caused by the magnitude of parameters in goal constraints. GP models (specifically (1.3), (1.5) and (1.6) in Chapter 1) that permit more than one goal constraint to be weighted or grouped at a single priority level run the risk of *incommensurability* of goal constraints. One way to avoid this problem is to put those goal constraints on the same comparable basis. By scaling to a common measure the of a_{ij} technological coefficients and b_i right-hand-side values, the magnitude of the parameters is adjusted to prevent incommensurability in the model. For example, suppose we have a weighted and preemptive GP model as in (2.5:

$$\text{Minimize: } Z = P_1 (d_1^- + d_1^+) + 0.1P_2 (d_2^- + d_2^+) + 0.9P_2 (d_3^- + d_3^+)$$

subject to:

$$x_1 + x_2 - d_1^+ + d_1^- = 100$$
$$1{,}000x_1 - d_2^+ + d_2^- = 100{,}000$$
$$x_2 - d_3^+ + d_3^- = 100$$

$$d_i^+, d_i^-, x_j \geq 0, \text{ for } i=1, \ldots, 3; \text{ for } j=1, \ldots, 2 \qquad (2.5)$$

It can be observed in this simple GP model that the values for the decision variables are not determined at the P_1 priority level. This goal only sets a boundary for the sum of the decision variables, such that $x_1 + x_2 = 100$. At P_2 the relative weighting is heavily in favor of making $x_2 = 100$ in the third goal constraint. Unfortunately the relative size of the parameters in the second goal constraint incorrectly causes the solution method seek to minimize the deviation in the second goal constraint, where the solution is obtained. The resulting solution for the (2.5) GP model is where $x_1 = 100$, $x_2 = 0$, $d_3^- = 100$, $P_1 = 0$, $P_2 = 90$ with the other deviations variables equaling zero.

To correct this problem (i.e., to make the relative weighting purposeful) Romero and Rehman (1984a) suggest a scaling of the goal constraint parameters. This can be accomplished by dividing the technological coefficients and the right-hand-side values by each goal constraint's respective right-hand-side value and multiplying the ratio by 100 to convert them into percentages of their objectives. So, the new coefficients would be found by:

New technological coefficient= (a_{ij} / b_i)x 100 (2.6)

New right-hand-side values= (b_i / b_i)x 100 (2.7)

Applying this logic to both goal constraints in the (2.5), the resulting revised GP model formulation is presented in (2.8). The resulting solution for (2.8) is where x_1 =0, x_2=100, d_2^- = 100, P_1 =0, P_2 =10 with the other deviation variables equaling zero. In this solution the more heavy weighting of 0.9 on the deviation variables in the third constraint has been included in the solution. Also the goal accomplishment at P_2 =10 is less because of the weighting, and therefore, we have a better solution or less deviation in the solution than in (2.5) where P_2 =90. Unfortunately, scaling to normalize parameters of differing goal constraints can make the interpretation of deviation more difficult and less meaningful. More over, other issues including *naive relative weighting* and *naive prioritization* can be created by changing the actual magnitude of GP model parameters. Other scaling methodologies have been reported in the literature to deal with some of these issues, such as *Euclidean norming* (Romero 1991, p. 40). For the best treatment of these subjects see Romero (1991, pp. 35-43). A review of the references in Table 2-4 is also recommended.

Minimize: $Z = P_1 (d_1^- + d_1^+) + 0.1P_2 (d_2^- + d_2^+) + 0.9P_2 (d_3^- + d_3^+)$

subject to:

$$x_1 + x_2 - d_1^+ + d_1^- = 100$$
$$x_1 - d_2^+ + d_2^- = 100$$
$$x_2 - d_3^+ + d_3^- = 100$$

$$d_i^+, d_i^-, x_j \geq 0, \text{ for } i=1, ..., 3; \text{ for } j=1, ..., 2 \qquad (2.8)$$

Analytic hierarchy process for determining goal weights and priorities: The *analytic hierarchy process* (AHP) is a statistical methodology (Saaty 1980). AHP can be used to derive relative weighting or priorities based on the weighting for GP models. In modeling situations where numerous qualitative factors are to be considered in establishing relative weighting, AHP may bring a recognized and uniformly fair framework for the assessment of those weights. For those interested in a basic review of the methodology, most introductory level MS/OR books now discuss this methodology (see Cook and Russell 1993, pp. 467-480). For other AHP methodologies and applications that might serve as a basis for combined GP/AHP paper ideas see Zahedi (1986). For specific applications of combined GP/AHP research see Olson, Venkataramanan and Mote (1986), Schniederjans and Wilson (1991), and Dyer, Forman and Mustafa (1992).

Conjoint analysis for determining relative weights or priorities: *Conjoint analysis* is a methodology that can be used to compute relative weighting or technological coefficients in GP models. Conjoint analysis has the ability to take large amounts of attributes that are measured using any type of measurement (i.e., rankings, rating scales, paired comparisons, etc.) and combine them into more a accurate parameter for use in a model. This methodology helps to minimize prediction error in parameters and permits reliability measures and validation statistics to be generated to help lend credibility to GP model structures. For a recent discussion on how conjoint analysis can be used in a variety of different areas see Green and Srinivasan (1990). For a GP model specific study using conjoint analysis see Srinivasan and Shocker (1973).

Regression analysis for determining relative weighting or goal constraint parameter estimation: GP in the form of a *constrained regression* model was used quite some time ago by Charnes, Cooper and Ferguson (1955). By minimizing deviation the GP model can generate decision variable values that are the same as the beta values in some types of multiple regression models. In Charnes, Cooper and Sueyoshi (1986, 1988) it was suggested that their GP model serves a valuable purpose of cross checking answers from other methodologies. Likewise, multiple regression models can also be used to more accurately combine multiple criteria measures that can be used in GP model parameters. Those parameters can include the relative weighting and the goal constraint parameters. The application of multiple regression suggested here is that in situations where the criteria is from the same population, a multiple regression model might be used to group the scores together to form more accurate and justifiable model parameters. For example, in project selection problems, managers often score projects on some scale (like 1 to 10 points). The managers then combine the points together to create a mathematical weighting or utility index on which to base their project selection decision. A multiple regression of the scores for each project will also generate a mathematical weighting that reflects their partial or weighted contribution of the project to some stated goal. The resulting beta coefficients from a regression model would be a more inclusive and in general, more scientifically derived than other simpler methods of estimation. In addition, all of the many beta statistical tests significance for purposes of reliability and significance can also be included in a research study to support the model's construction. One application of the use of regression in estimating goal constraint parameters is in Kwak, Schniederjans and Warkentin (1991).

Logarithmic transformations of goals: In a variety of problem situations the goal constraints must be expressed an nonlinear functions (see Philipson and Ravindran (1978, 1979), Singh (1983) and Sundaram (1978)). We will be discussing *nonlinear GP* in Chapter 3. There is nothing wrong with structuring GP models with nonlinear goal constraints. Unfortunately, such constraints can add complexity to the solution process, which in turn leads to model formulation issues (i.e., dominance, etc.). To avoid the complications of dealing with goals expressed as polynomials many researchers convert the nonlinear functions to linear equivalents using a *logarithmic transformation* (see Romero 1991, pp. 67-71). This can permit a standard linear GP software package to be used to obtain a solution from the model. Romero and Amador (1986) pointed out that such transformations can yield incorrect results because of the unique nature of GP models. They suggested that logarithmic transformations can be used correctly if applied to LP constraints. This leaves open the option of converting goal constraints back into LP constraints for purposes of dealing with the nonlinearity. Where this is not possible, Romero and Amador (1986) suggest that if the logarithmically transformed goal constraints are placed at the P_1 priority level and possess no deviation in their resulting solution (i.e., a fully achieved goal), then the logarithmic transformation is licit.

Input-output analysis for technological parameter estimation: The use of the economics based methodology called *input-output analysis* to generate technological coefficients can be traced to LP model applications by Ijiri (1968), Gambling (1968) and Livingston (1969). What input-output analysis can be used to do in GP models is to transform technological coefficients to reflect real world impact of different types of *posterior* system behavior. For example, in any production process some units of material resources become scrap. If scrap is an important factor in a decision making environment it is necessary to include such allowances in a GP model. In highly complex multi-stage production processes where the flow of production moves in-and-out of various production processes, a simple percentage scrap rate deduction in technological coefficients reflecting usage by a particular production process, generally will not provide a realistic posterior output parameter for the GP model. Input-output analysis, using an inverse matrix procedure, adjusts technological coefficients to take into consideration the interaction and resulting posterior system behavior between all production process. Input-output analysis methodology is illustrated in Kendall and Schniederjans (1983) where a hypothetical example of how this methodology works in a GP model is presented. A real world example of the use of input-output analysis in a GP model of a manufacturing operation is presented in Schniederjans and Markland (1986).

There are almost as many additional tactical methodologies used to improve GP models as there are papers not cited in this section. Some of these methodologies will be discussed in Chapter 3 where the solution method used to generate an answer from a GP model can impact the formulation of that model.

SUMMARY

This chapter is about formulating GP models. It began with a simple GP model formulation procedure. Recognizing that the procedure is not always applied correctly, this chapter discussed issues that can lead to a GP model being misformulated.. Such issues as dominance, inferiority and inefficiency in GP solutions were discussed along with issues of naive relative weighting, incommensurability, naive prioritization and redundancy. Having stated what to avoid or not to do in formulating GP models, the last section of this chapter was devoted to ideas for formulation strategies and tactics that could be used to improve GP models.

The same people who formulate GP models are also the same people who review GP models for publication in journals. It is hoped that this chapter will help the GP researcher who formulates GP models to better understand how to formulate a GP model. It is also hoped that this chapter will help the reviewers of GP models understand that issues raised in the literature that are used to reject GP models for publication purposes may themselves be in error in a particular decision environment. That care by both model builders and model critics is a necessary requirement for the perpetuation of the *life cycle of GP research.*

REFERENCES

All references in this chapter, except those below, can be found in *Appendix B, Journal Research Publications on Goal Programming.*

Charnes, A. and Cooper, W. W., *Management Models and Industrial Applications of Linear Programming*, Vols. 1 & 2, John Wiley and Sons, New York, NY, 1961.

Cohon, J. L., *Multiobjective Programming and Planning*, Academic Press, New York, NY, 1978.

Cook, T. M. and Russell, R. A., *Introduction to Management Science*, Prentice Hall, Englewood Cliffs, NJ, 1993.

Fang, S. and Puthenpura, S., *Linear Optimization and Extensions: Theory and Algorithms*, Prentice Hall, Englewood Cliffs, NJ, 1993.

Gambling, T. E., "A Technological Model for Use in Input-Output Analysis and Cost Accounting," *Management Accounting*, Vol. 15 (1968), pp. 33-38.

Green, P. E. and Srinivasan, V., "Conjoint Analysis in Marketing: New Developments with Implications for Research and Practice," *Journal of Marketing*, Vol. 54, No. 4 (1990), pp. 3-19.

Hillier, F. S. and Lieberman, G. J., *Introduction to Operations Research*, 4th ed., Holden-Day, Oakland, CA, 1986.

Ignizio, J. P., *Introduction to Expert Systems*, McGraw-Hill, New York, NY, 1991.

Ignizio, J. P., *Linear Programming in Single- and Multiple-Objective Systems*, Prentice-Hall, Englewood Cliffs, NJ, 1982.

Ijiri, Y., "An Application of Input-Output Analysis to Some Problems in Cost Accounting," *Management Accounting*, Vol. 15 (1968), pp. 49-61.

Kwak, N. K. and Schniederjans, M. J., *Introduction to Mathematical Programming*, Krieger Publishing, Malabar, FL, 1987.

Livingston, J. L., "Input-Output Analysis for Cost Accounting, Planning, and Control," *The Accounting Review*, Vol. 21 (1969), pp. 48-64.

Pareto, V., *Manual of Political Economy*, A. M. Kelley, New York, NY, 1971.

Romero, Carlos, *Handbook of Critical Issues in Goal Programming*, Pergamon Press, Oxford, England, 1991.

Saaty, T. L., *The Analytic Hierarchy Process*, McGraw-Hill, New York, NY, 1980.

Schniederjans, M. J., *Linear Goal Programming*, Petrocelli Books, Princeton, NJ, 1984.

Simon, H. A., "A Behavioral Model of Rational Choice," *Quarterly Journal of Economics*, Vol. 69 (1955), pp. 99-118.

Simon, H. A., *Models of Man*, John Wiley and Sons, New York, NY, 1957.

Zahedi, F., "The Analytic Hierarchy Process-A Survey of the Methods and Its Applications," *Interfaces*, Vol. 16, No. 4 (July-August 1986), pp. 96-108.

CHAPTER 3. GOAL PROGRAMMING SOLUTION METHODOLOGY

When goal programming (GP) was introduced in the mid 1950's there was little computer software (or computers) to help support the growth of this computationally dependent methodology. By the early 1970's both computers and software applications where in place to encourage the development of GP modeling. As the *life cycle of GP research* in Figure 1-3 in Chapter 1 reveals, the real growth period for GP literature began at this very same time. The GP software required the availability of GP algorithms used to generate the primary GP problem solutions. In addition, a collection of supporting algorithms are also necessary to permit a post-solution analysis or secondary consideration of the solutions obtained in the primary solution. Collectively, these *primary* and *secondary* algorithms can be called *GP solution methodologies.*

The purpose of this chapter is to review all of the various types of GP solution methodologies that have appeared in the literature. This review includes the primary GP algorithms and methodology used to generate linear GP, integer GP and nonlinear GP solutions. In addition, secondary GP methodologies including duality and sensitivity analysis used to obtain post-solution information will also be discussed.

PRIMARY GOAL PROGRAMMING SOLUTION METHODOLOGIES

There are many different methodologies and algorithms used to generate solutions for GP models. We will begin by categorizing them into four groups of *linear GP* (which includes all linear based GP solution methods), *integer GP* (which includes methodology used to generate all integer, mixed integer and zero-one integer solutions), *nonlinear GP* (which includes all nonlinear based GP solution methods), and a final *other* group for all methodology that does not fit into the other three groups.

It should be mentioned that applications of GP methodologies add to GP solution methodologies. Some of the methodological improvements that are a part of GP literature have been presented as an application or in a case study. These methodologies will not be included in this chapter since they are presented as applications. These applications will be discussed in Chapter 4 and readers are encouraged to review that chapter's citations of applications to complete their survey of any one type of methodology introduced in this chapter.

Linear Goal Programming Algorithms and Methodology

The first linear GP algorithm is actually an LP algorithm. The methodological proof for solving LP models structured as GP problems can be found in Charnes and Cooper (1961, pp. 210-215). With the improvements of preemption, the *generalized inverse approach* and the illustrative use of the *simplex based algorithm* by Ijiri (1965), as well as the publication of a software program by Lee (1972), substantially increased linear GP research in methodological improvements. While it was assumed the LP proof by Charnes and Cooper (1961) was sufficient to justify the mathematical workings of GP algorithms, it is interesting to note that no mathematical proof of a simplex-based linear GP methodology actually appeared until Evans and Steuer (1973).

In Chapter 2 the distinctions between weighted GP models, preemptive GP models and various combinations were discussed. While some GP algorithms can only be used with a single type of GP model, others have been designed to handle a wider variety of GP models. This logic has been taken to the extreme in MULTIPLEX model and algorithm (Ignizio 1985c), which claims to be able to work with LP, weighted GP, preemptive GP and fuzzy GP models.

The basic algorithms used to solve the weighted GP, preemptive GP and their combinations are available in virtually all basic *management science/operations research* (MS/OR) books (see typical example in Turban and Meredith 1991, Chapter 9). A more detailed discussion of such algorithms can be found in GP books, including Ignizio (1976,1982), Ijiri (1965), Lee (1972) and Schniederjans (1984). Other extensions of methodology can be found in Table 3-1.

Many of the methodological refinements used in weighted GP and preemptive GP models, can be found in applications of these models. These applications are listed in Chapter 4.

Table 3-1. Citations on Weighted/Preemptive GP Methodology

Reference	What Reference Provides
Arthur and Ravindran (1978), Schniederjans and Kwak (1982), Kwak and Schniederjans (1982, 1985c)	reduced size algorithms
Freed and GLover (1981a, 1981b)	used as discriminat analysis
Charnes and Cooper (1977), Evans and Steuer (1973), Hwang, Paidy, Yoon and Masud (1980)	mathematical proofs for GP
Schenkerman (1991)	discussion on weighted GP
Knoll and Engelberg (1978),Kluyver (1979), Sherali (1982),Shim and Siegel (1975), Spivey and Tamura (1970), Steuer (1979), Widhelm (1981)	weighted GP methodologies
Arthur and Ravindran (1980), Charnes and Gooper (1961), Dauer and Krueger (1979), Ignizio (1976 book, 1982 book, 1985c), Iserman (1982), Ijiri (1965 book), Lee (1972 book), Schniederjans (1984 book)	algorithms for both models
Lee (1983), Lee and Rho (1979b, 1983, 1986)	decomposition methodologies
Crowder and Sposito (1987), Ignizio (1985a, 1987)	solution by dual solution
Akgul (1984), Alvord (1983), Clayton and Moore (1972b), Gibbs (1973), Hindelang (1973), Ignizio (1978a, 1983a), Ruefi (1971)	general discussion of issues

Integer Linear Goal Programming Algorithms Methodology

In GP problem situations where decision variables are restricted to integer values, special integer GP methodologies were developed. Most of the GP methodologies are based on integer LP methodologies. For example, in all or mixed integer LP problems one of the most common integer methodologies is the *branch-and-bound solution method.* Arthur and Ravindran (1980) developed their branch-and-bound *integer GP* algorithm on this same LP algorithm. Additional citations for these methods can be found in Table 3-2. In the case of zero-one LP integer solutions the most commonly approach is some type of *enumeration method.* Garrod and Moores (1978) developed their *zero-one GP* solution methodology using the same approach. Other citations for zero-one GP methods can be found in Table 3-3.

It is interesting to note that most of the research in integer GP is applied. In Chapter 4 dozens of application papers are presented using methodologies based on one that is presented in this section. A review of these applied models will reveal additional methodological refinements that are not included in the studies cited in this section.

Nonlinear Goal Programming Algorithms and Methodology

According to Saber and Ravindran (1993) there are four major approaches to nonlinear GP: (1) *simplex based nonliner GP*, (2) *direct search based nonlinear GP*, (3) *gradient search based nonlinear GP*, and (4) *interactive approaches to nonlinear GP*. We will discuss each of these in this section. For additional information on these subjects review the citations presented in Table 3-4.

The *simplex based nonlinear GP* approaches include the method of *approximation programming*, which was developed by Griffith and Steward (1961) adapted by Ignizio (1976) for GP. This methodology permits nonlinear goal constraints to be included in a GP model.

Another simplex based approach to nonlinear GP is called *separable programming*. Originally developed by Miller (1963), this approach was modified for GP by Wynne (1978). This methodology allows nonlinear goal constraints to be included in the GP model by restricting the range of the decision variables into separable functions that are assumed linear. This methodology is based on the logic of *piece-wise linear approximations.* For a review of the mechanics of this methodology see Reklaitis, Ravindran and Ragsdell (1983).

Table 3-2. Citations on Pure/Mixed Integer GP Methodology

Reference	Notes and Comments on What Reference Provides
Arthur and Ravindran (1980)	introduction of branch-and-bound method
Sharif and Agarwal (1976), Lee and Morris (1977)	classical introductions to the subject
Gabbani and Magazine (1986)	a combined methodology with heuristics
Gupta and Sharna (1989)	a combined methodology with quadratic programming
Hallefjord and Jornstern (1988), Ignizio (1989), Sharma and Sharma (1980)	all discuss the pro's and con's of integer programming
Ignizio and Daniels (1985)	a combined methodology with fuzzy GP
Ignizio (1983c, 1984), Ignizio (1985b), Markland and Vickery (1986)	all new methodologies

Table 3-3. Citations on Zero-One GP Methodology

Reference	Notes and Comments on What Reference Provides
Garrod and Moores (1978)	an implicit enumeration method
Lee and Luebbe (1988)	a comparative evaluation of a variety of methods
Lee and Luebbe (1987b)	a methodology utilizing a partitioning method
Rasmussen (1986), Wilson and Jain (1988), Gen, Ida and Lee (1990)	all provide new methodologies or innovations

Table 3-4. Citations on Nonlinear GP Methodology

Reference	Notes and Comments on What Reference Provides
Armstrong, Charnes, and Haksever (1988), Weistoffer (1983)	methodology combined with interactive procedure
Awerbuch, Ecker and Wallace (1976), Charnes, Cooper, Klingman and Niehaus (1975)	a general discussion and comments
Clayton, Weber, and Taylor (1982)	methodology combined with simulation methods to generate a more greedy or efficient solution
Davis and Whitford (1985)	comments on decomposition method

Table 3-4. (Continued)

Reference	Notes and Comments on What Reference Provides
Deckro and Hebert (1988) El-Sayed, Ridgely, Sandgren (1989)	method for polynomial preference structure analysis
Dinklebach (1967)	methodology combined with fractional GP
El-Dash and Mohamed (1992)	methodology combined duality and sequential GP
Gupta and Sharma (1989), Reeves (1978), Shim and Siegal (1975)	methodologies using quadratic GP
Lee (1985a)	methodology combined with chance constraints
Lee and Olson (1985)	methodology combined with chance constraints
Lee and Lerro (1973)	methodology combined with calculus
Masud and Hwang (1989), Shephead(1980)	extensions of basic methodologies
Olson (1992)	review of methodologies
Romero and Amador (1986)	a discussion on nonlinear transformations
Saber and Ravindran (1992)	methodology based on partitioning
Saber and Ravindran (1993)	a survey of methodology
Sakawa (1985)	methodology combined with interactive and fuzzy GP

Still another simplex based approach to nonlinear GP is *quadratic goal programming*. An early discussion of quadratic programming methodology can be found in Beale (1967). The idea of this approach as a GP methodology was originally presented by Gary Reeves at the 1977 Joint Annual Conference of the Operations Research Society and the Institute of Management Science in San Francisco. Quadratic GP permits quadratic goal constraints and quadratic deviation variables in the objective function. For a review of the mechanics of this methodology see Ringuest and Gulledge (1982) and Gupta and Sharma (1989).

Direct search based nonlinear GP methods utilize some type of logical search pattern or methods to obtain a solution that may or may not be the best satisfising solution. The logic process is based on repeated attempts to improve a given solution by evaluating its objective function and/or goal constraints. The basic search idea originated with Box (1965), but was applied to GP by many others including Nanda, Kothari and Lingamurthy (1988). There are many search methods in the literature. Hooke and Jeeves (1961) developed a single objective, continuous variable, unconstrained optimization method, that was later adapted for GP by Ignizio (1976) and Hwang and Masud (1979). An alternative approach that utilizes simulation methods in the search procedure is presented by Clayton, Weber and Taylor (1982).

Gradient based nonlinear GP methods use calculus or partial derivatives of the nonlinear goal constraints or the objective function to determine the direction in which the algorithm is to search for a solution and the amount of movement necessary to achieve that solution. While gradient based methods are generally more efficient in obtaining a solution, they may not be appropriate for GP models whose goal constraints or objective function is nondifferentiable.

Based on Zoutendijk's (1960) algorithm Lee (1985a), Lee and Olson (1985), Newton (1985) and Olson and Swenseth (1987) all developed a version of the gradient method for GP called the *chance constraint method*. The chance constraint method allows parameters to be distributed along a probability distribution. The introduction of the probability distribution is where this methodology obtained its probabilistic or chance name. The use of the chance constraint method requires the assumption that the technological coefficients are normally distributed. There are a dozen applications of chance constraint GP for linear and nonlinear problems listing in Chapter 4. Some of these date back to the 1970's and range up to the 1990's.

Another gradient based method is called the *partitioning gradient method*. Developed for linear GP by Arthur and Ravindran (1978) using a simplex based approach, this methodology can be highly efficient in obtaining nonlinear GP solutions. It works on the basis of finding smaller subproblems that lead to an optimal solution. By solving these smaller problems and eliminating decision variables from the model, the size of the model is reduced. For information on a software application and the methodology itself, see Saber (1991). A special version of the gradient based method is called the *decomposition method*. The decomposition method can solve linear GP or nonlinear GP problems. It is usually based on some version of the Dantzig and Wolfe (1960) LP decomposition method, where large models are decomposed into smaller submodels whose solution will be used to generate the solution to the original larger model. Algorithms and research on the decomposition method for GP can be found in , Ruefi (1971), Sweeney, Winkafsky, Roy and Baker (1978), Lee (1983), and Lee and Rho (1979a, 1979b, 1985, 1986).

One special type of nonlinear GP methodology can be called *stochastic goal programming*. In a stochastic GP model there are probability distributions present to describe model parameters or the model's structure. The chance constraint methodology for example can be used to model and solve some classes of stochastic GP problems. For a good review of the basics see Contini (1968). Other extensions of methodology can be found in Table 3-5.

Interactive approaches to nonlinear GP or *interactive GP* can be defined as a collection of methodologies that are based on progressive articulation of a decision maker's preferences in a decision environment (see Dyer 1972, Benson 1975). The decision maker using interactive GP will be lead to a better solution by interactively comparing a given solution. This makes interactive GP a sequential search process, but one that involves periodic feedback to the decision maker to guide the direction of the search. The term *sequential GP* is often used with the interactive approach to better describe the step-wise nature of this methodology. For additional methodological sources in sequential GP see Table 3-6.

Table 3-5. Citations on Stochastic GP Methodology

Reference	Notes and Comments on What Reference Provides
Ben-Tal and Teboulle (1986)	a discussion on utility, penalty functions and duality
Contini (1968)	classic introduction to the subject
Hussein (1993)	combined methodology with an interative GP approach
Ketzloff-Roberts and Morey (1993)	methodology combined with data envelopment analysis
Odom, Shannon and Buckles (1979)	a commentary of the subject
Reznicek-Roberts and Morey (1993)	combined methodology with data envelopment analysis
Sengupta (1981)	basic review of methodology
Stancu-Minasian and Tigan (1988)	methodology combined with fractional GP
Sueyoshi (1991)	methodology combined with data envelopment analysis
Teghem, Dugrane, Thauvoye and Kunsch (1986)	combined methodology with interative GP

Interative GP has been used for all types of GP models (i.e., linear GP, integer GP and nonlinear GP). Any of the methods that are used to solve GP problems can be used as an interactive, sequential GP search methodology. The combination of these terms appeared in Hasud and Hwang (1981), which may have lead to the often used combination. For an excellent review if the mechanics of the various methods see Van Delft and Nijkamp (1977), Spronk (1981). Additional citations on sequential GP appear on Table 3-7.

Table 3-6. Citations on Interactive GP Methodology

Reference	Notes and Comments on What Reference Provides
Buchanan and Daellenbach (1987)	a comparative evaluation of a variety of methods
DeKluyver and Moskowitz (1984)	a combined methodology to estimate probabilities in forecasting
Dyer (1972), Masud and Hwang (1981), Walker (1978), Zoints and Wallenius (1976)	classic introductions to the subject
Ferreira and Geromel (1990), Fichefit (1976), Gabbani and Magazine (1986), Gass and Dror (1983), Korhonen and Laakso (1986), Reevesand Hedin (1993), Steward (1988) Teghem, Dugrane,Thauvoye and Kunsch (1986) Weistroffer (1982, 1983), Yang, Chen and Zhang (1990)	all provide new methodologies or innovations
Gibson, Bernardo, Chung and Badinelli (1987)	a comparison of various methods
Rustangi and Bare (1987)	discusses how to resolve goal conflicts
Sakawa (1985)	a combined methodology with fuzzy and nonlinear GP

Table 3-6. (Continued)

Reference	Notes and Comments on What Reference Provides
Sakawa and Gen(1985), Sasaki, Gen and Ida (1990)	combined methodology with fuzzy and sequential GP
Shin and Ravindran (1991)	a survey of methods
Tingley and Liebman (1984b)	discusses how to adjust parameters in model

Table 3-7. Citations on Sequential GP Methodology

Reference	Notes and Comments on What Reference Provides
Crowder and Sposite (1991)	usage with a specific computer application
El-Dash and Mohamed (1992), Markowski and Ignizio (1983a)	a combined methodology with duality
Ignizio (1979)	usage with a specific computer application
Masud and Hwang (1981)	an introduction to the basics
Ogryczak (1988a)	a unique discussion and methodology
Saaki and Gen (1992)	a combined methodology with fuzzy and interactive GP

Other GP Algorithms and Methodology

There are at least four other algorithm based methodologies that are extensively represented in GP literature: *interval GP*, *fractional GP*, *duality solution* and *fuzzy GP*. Each of these other methodologies can and often are used with linear GP, integer GP and nonlinear GP models. They also offer unique modeling features that have distinguish them in their own right.

Interval GP: Interval GP allows parameters, particularly the right-hand-side goal values to be expressed on an interval basis. This method is based on interval LP, where an upper boundary, b_u and lower boundary, b_l for the right-hand-side values can be stated as:

$$b_l \leq a_{ij}\, x_j \leq b_u \tag{3.1}$$

So the interval GP equivalence would be accomplished with two goal constraints:

$$a_{ij}\, x_j - d_u^+ + d_u^- = b_u \tag{3.2}$$

$$a_{ij}\, x_j - d_l^+ + d_l^- = b_l \tag{3.3}$$

Where the d_u^+ and d_l^- are both minimized in the objective function and the other deviation variables are free to permit some compromised value for the resulting right-hand-side value. This method can be used to deal with a variety of formulation issues that are used to criticize GP models, such as the inappropriateness of predetermined goals or targets (see Min and Storbeck 1991). For citations on interval GP methodology see Table 3-8.

Fractional GP: Fractional GP is a methodology used when modeling ratios. In a variety of situations, such as modeling return on investment problems, market share problems or percentage type problems, fractional GP maybe the most appropriate of the GP methodologies. As Awerbuch, Ecker and Wallace (1976) noted, there complexities in GP model formulations that make simple multiplication of goal constraints an invalid means for dealing with fractional values. For a review of some of the controversy see Hannan (1977, 1981) and Soyster and Lev (1978). Fractional GP is also an extension of LP, called *fractional LP* (see Marto (1964), Bitran and Novaes (1973). For citations on interval GP methodology see Table 3-9.

Table 3-8. Citations on Interval GP Methodology

Reference	Notes and Comments on What Reference Provides
Charnes and Collomb (1972)	a classic introduction to the basics
Charnes and Cooper (1977)	a mathematical proof and explanation
Ichida and Fujii (1990)	a basic methodological example
Inuiguchi and Yosufumi (1991)	a methodological study of interval parameters
Ishibuchi and Tanaka (1990)	a methodological study of an interval objective function
Steuer (1979)	a combined methodology with sensitivity analysis for weighting

Duality Solution: It has been shown that GP models can be solved more efficiently and without some computational problems by solving the dual formulation of the a GP model (Dauer and Krueger 1977, Ignizio 1985). This method is not without its problems as observed by Crowder and Sposito (1987) and replied to by Ignizio (1987). An interesting extension of this method to sequential nonlinear GP can be seen in El-Dash and Mohamed (1992).

Fuzzy GP: Fuzzy GP is based on *fuzzy set theory*. Fuzzy sets are used to describe imprecise goals. These goals are usually associated with objective functions and are used to reflect both a weighting (with values from zero to one) and range of goal achievement possibilities. For example, if a profit function ranges from \$100 to \$300 in increments of \$100, we might weight our preference for \$100 occurring as 0.0, \$200 occurring 0.5 and \$300 occurring as 1.0. The numerical relationship between the goal of profit and the weighting attached to them, is a fuzzy set of numbers defining the decision makers utility

Table 3-9. Citations on Fractional GP Methodology

Reference	Notes and Comments on What Reference Provides
Agrawal, Swarup, and Garg (1984), Armstrong, Charnes, and Haksever (1987)	basic discussion and select methodology
Awerbuch, Ecker and Wallace (1976), Charnes and Cooper (1962), Hannan (1977, 1981b), Joksh (1964), Kornbluth and Steuer (1981a, 1981b), Soyster and Lev (1978)	all are classical introductions and commentary
Dinklebach (1967)	a combined methodology with nonlinear GP
Lee, Chung, and Tcha (1991)	a combined methodology with fuzzy GP and duality
Pant and Shah (1992)	a new linear approach
Stanu-Minasian and Tigan (1988)	a combined methodology with stochastic GP

in the profit occurrences. The relationship between the weighting and the profit function can be linear or nonlinear. Most importantly, this methodology allows the decision maker who can not precisely define goals to at least express them using a weighting structure that is not limited. This makes fuzzy programming an idea approach when utility function type goals are to be used in the GP model. For citations on fuzzy GP methodology see Table 3-10.

Table 3-10. Citations on Fuzzy GP Methodology

Reference	Notes and Comments on What Reference Provides
Carlsson (1982), Ignizio and Hannan (1982), Llena (1985), Mohandos, Phelps and Ragsdell (1990), Narasimhan (1981)	all provide commentary and criticism of the subject
Hannan (1982a)	explains the difference between fuzzy GP and other fuzzy methodologies
Hannan (1981a, 1981c, 1981d), Narasimhan (1980), Zimmerman (1978), Zimmerman (1983)	all provide basic introductions to the subject
Ignizio and Daniels (1984)	a combined methodology with integer GP
Lee, Chung, and Tchu (1991)	a combined methodology with fractional GP
Rao, Tiwari and Chakraborty (1993)	a combined methodology with chance constraint GP
Rao, Tiwari and Mohanty (1988a, 1988b)	a preference structure analyses
Rubin and Narasimhan (1984)	a combined methodology with Nester priorities
Sakawa (1985)	a combined methodology with interactive and nonlinear GP

Table 3-10. (Continued)

Reference	Notes and Comments on What Reference Provides
Sasaki and Gen (1992)	a combined methodology with interactive GP
Tiwari, Dharmar, and Rao (1986)	a priority structure analysis
Tiwari, Dharmar, and Rao (1987)	an additive model methodology

SECONDARY GOAL PROGRAMMING SOLUTION METHODOLOGIES

Two extensions of LP are duality and sensitivity analysis. These extensions exist in GP as well, but with some unique characteristics.

Duality in GP

In LP models we seek to determine the *marginal contribution* (also called the *dual decision variable*) of each of the right-hand-side values in terms of the single objective function units (Fang and Puthenpura 1993, pp. 56-72). The same basic simplex process is used in GP duality to derive the marginal contribution of each right-hand-side values. A variety of GP concepts and methodologies on duality can be found in Markowski and Ignizio (1983a, 1983b), Ogryczak (1986, 1988b) and Martinez-Legaz (1988). An exception that makes GP duality different is that its interpretation of the resulting marginal contribution is some what different from LP. The marginal contribution of right-hand-side values or goals in GP models take on a *multidimensional characteristic* (Ignizio 1984b). The interpretation of the marginal contribution in GP models has to be in terms of all of other goals in the model. That is, the marginal contribution of one goal in terms of all other goals. An excellent discussion of the mechanics and interpretations can be found in Ignizio (1982, Chapter 18).

Other studies have extended duality analysis in GP. An iterative algorithm with its dual formulation was discussed by Dauer and Krueger (1977). Ben-Tal and Teboulle (1986) added an even greater degree of complexity to the use of duality with a stochastic, nonlinear GP model. Likewise, Lee, Chung and Tcha (1991) combined duality in GP with fuzzy GP and fractional GP. As previously mentioned, dual formulations for GP models have also been shown to enhance computational efficiency for solving GP problems when compared to other standard algorithms (see Dauer and Krueger 1977, Ignizio 1985).

Sensitivity Analysis in GP

According to Ignizio (1982, Chapter 19) there are seven types of changes that can be implemented as a part of sensitivity analysis in GP: (1) changes in the weighting at a priority level, (2) changes in the weighting of deviation variables within a priority level, (3) changes in right-hand-side values, (4) changes in technological coefficients, (5) changes in the number of goals, (6) changes in the number of decision variables, and (7) reordering preemptive priorities. Most of these have been illustrated by application (see Chapter 4). Lee (1972, Chapter 7), Ignizio (1982, Chapter 19), Schniederjans (1984, pp. 105-109.) all provide basic methodologies for undertaking these seven types of sensitivity analyses in GP models. For additional citations on GP sensitivity analysis methodology see Table 3-11.

COMPUTER SOFTWARE SUPPORTING GP SOLUTION ANALYSIS

In S. M. Lee's 1972 GP book, *Goal Programming for Decision Analysis*, the computer coding for a FORTRAN program presented the first published source of software for all the various types of weighted and preemptive linear GP models. The availability of this code and the interest in GP in the early 1970's undoubtedly lead to the greatest growth stage in GP research that this subject has ever had (see Figure 1-3 *Life Cycle of GP Research*). Based on citation counts, Lee's software or versions developed from it are the mostly cited software used to solve GP models during the 1970's and early 1980's. Many of the GP software systems used in the 1990's are based on a version of Lee's GP software program (see Lee, 1972, pp. 140-157). Other mainframe computer based systems, like Ignizio's (1985c) MULTIPLEX code that came later helped to broaden software capabilities to included LP algorithms as a part of a package of software. Being used for mainframe computer systems, both Lee's and Ignizio's software applications could be reprogrammed to handle models equal to the memory of the computer on which they were run.

Table 3-11. Citations on Sensitivity Analysis GP Methodology

Reference	Notes and Comments on What Reference Provides
Gambicki and Haimes (1975)	a basic methodology for changes in goals
Karandikar and Farrokh (1987)	a unique solution methodology
Rifai (1980)	a basic methodology for changes in goals
Shim and Siegel (1980)	a methodology for changes in priorities
Steuer (1979)	a methodology for changes in weighting
Wilson and Jain (1988)	a methodology for use with zero-one GP

Other specialized computer codes whose ability to deal with a smaller subset of GP problem solving have been developed over the years. Unfortunately, most such codes do not end up in journal publications and even their applications are not always reported until years after they appear in the literature. For example a nonlinear GP code called the *Partitioning Gradient Based* (PGB) computer program appeared in Lasdon, Waren, Jain and Ratner (1978). Its coding application to nonlinear GP occurred much later by Saber (1991).

For the purposes of this book, the AB:QM, Version 3.1 (Lee 1993) microcomputer or PC was more than adequate to do solve the small problems presented. The AB:QM software can handle a 50 goal constraint by 50 decision variable or (50 row x 50 column) GP model. It does not handle integer GP problems or nonlinear GP models, unless those models can be converted into the linear GP equivalent. AB:QM also does not provide duality or sensitivity analysis for GP models. This package, like most of those sold with MS/OR books is designed to handle classroom problems for teaching purposes using a wide variety of quantitative methods. While some small GP applications can work within the limitations of this type of microcomputer software, real world applications need much more GP algorithm power.

A telephone survey of existing GP software available through US developers was undertaken in the Fall of 1994 for this book. The purpose of this survey was to locate as many GP software applications for both microcomputers and mainframe computers as were available at that time of the survey. Excluding the MS/OR book software applications, the focus of this survey was on finding packages that would support research and real world sized GP models. A review of other software surveys on MS/OR methodology was used as a baseline for initial survey software developer contacts (Sharda 1992; Saltzman 1994). Other software developer contacts were obtained from advertisements in computer publications and from actual citations in GP articles.

Out of a total of 112 identified developers of LP or some type of mathematical programming software application, only 15 developers actually claimed that GP type models could be processed by their software. The respondents of this survey are presented in Table 3-12. The survey sought to determine the computer system platform or hardware on which the software worked. Platforms included *microcomputer systems*, *mainframe computer systems* or both. The capacity or size of GP models that could be processed by the various computer applications was also determined. In this regard it should be noted that when the word *memory* is used it refers to the size of the individual computer's memory capacity. This feature simply means the software is flexible to adapt to variable computer memory space limitations. In cases where developers offered a variety of packages capable of dealing with a range of differing sized models, only the largest size is reported in Table 3-12. Most importantly, the survey sought to determine which of the various GP methodologies were available on the software applications. The following six options were *linear GP*, *integer GP*, *nonlinear GP*, *duality*, *sensitivity analysis*, and *other*. If any of these six options were present in any form for the computer software application, a *Yes* appears in Table 3-12, other wise a *No* appears. It should be noted that in some cases the solution methodology used to generate GP solutions is based on LP algorithms used in a *sequential* manner to achieve the GP desired solution results.

Like all surveys, this one has some limitations. One limitation is the recognition that more then just those respondents is this survey offer GP software. For example, developers in nations other than the US were not surveyed or included in this study because of cost and time factors. Another limitation on this survey is the inherent bias that comes from accepting information based on the telephone responses of the representatives of the software developers, rather than actual experience with each software system. The claims of software capabilities are what were reported in the telephone survey by representatives of the firms they represent and may not prove out in the actual use of the specific software reported in this survey. In other cases, like with GINO (see Liebman, Lasdon, Schrage and Waren 1986), a general purpose modeling language, the software requires the GP model to be expressed in a slightly different form than those presented in the GP literature. This extra modeling effort may inhibit some applications informational efficacy.

Even given these limitations on a less the complete sampling of computer packages, the results reveal both quantity and quality of GP algorithm support in existing GP software applications. There appears to be more than enough GP algorithm power to solve any type of GP problem situation. Indeed, some of the newer software applications are particularly powerful when combined together. For example, the GAMS and MINOS software systems can be jointed together to overcome their individual limitations (see Brooke, Kendrick and Meeraus 1988). Combined the GAMS/MINOS software application can handle a larger range of problems and provides addition solution analysis information. An interesting capability of a combined software system is illustrated by MINOS and CONOPT, both of which are produced by the same software developer. CONOPT is not a GP software application but adds modeling diagnostic capability to MINOS GP models, such that CONOPT could help diagnose model formulation errors in GP models. Such software combinations are endless in variety and are left to the experimenters to research and report in the literature.

Table 3-12. Computer Software Applications That Support GP Solution Analysis

Software:	AMPL	CPLEX Mixed Integer Optim.	GAMS
Publisher:	Boyd & Fraser Pub. One Corp. Place Ferncroft Village Danvers, MA 01923	CPLEX Optim. Inc. 930 Tahoe Blvd. #802-279 Incline Village, NV 89451	Boyd & Fraser Pub. One Corp. Place Ferncroft Village Danvers, MA 01923
Phone No.:	(508)777-9069	(702)831-7744	(508)777-9069
FAX No.:	(508)777-9068	(702)831-7755	(508)777-9068
System Platforms:	Micro, Mainframe	Micro, Mainframe	Micro, Mainframe
Size Capacity (rows x cols.):	Memory	Memory	Memory
System Features:			
Linear GP	Yes	Yes	Yes
Integer GP	Yes	Yes	Yes
Nonlinear GP	Yes	Yes	Yes
Duality	Yes	Yes	No
Sens. Analy.	Yes	Yes	Yes
Other		Diagnostics	

Table 3-12. (Continued)

Software:	Extended LINDO	Extended GINO	HS/LP
Publisher:	LINDO Systems 1415 N. Dayton Str. Chicago, IL 60622	LINDO Systems 1415 N. Dayton Str. Chicago, IL 60622	Haverly Syst. Inc. P. O. Box 919 Denville, NJ 07834
Phone No.:	(312) 871-2524	(312) 871-2524	(201)627-1424
FAX No.:	(312) 871-1777	(312) 871-1777	(201)625-2296
System Platforms:	Micro, Mainframe	Micro, Mainframe	Micro
Size Capacity (rows x cols.):	32,000x100,000	800x1,600	8,192xMemory
System Features:			
Linear GP	Yes	Yes	Yes
Integer GP	Yes	No	Yes
Nonlinear GP	No	Yes	Yes
Duality	Yes	Yes	No
Sens. Analy.	Yes	Yes	Yes
Other			

Table 3-12. (Continued)

Software:	IBM Optimization Subroutine Library	MINOS, NPSOL and LSSOL	MPSX-MIP/370
Publisher:	IBM Dept. 85BA, MS 658 Neighborhood Rd. Kingston, NY 12401	Stanford Business Software Inc. 2672 Bayshore Pkwy. Mtn. View, CA 94043	Altium, of IBM IBM MS 936 Neighborhood Rd. Kingston, NY 12401
Phone No.:	(914)385-5027	(415)962-8719	(914)385-6408
FAX No.:	(914)383-4239	(415)962-1869	(914)385-4500
System Platforms:	Micro, Mainframe	Micro, Mainframe	Micro, Mainframe
Size Capacity (rows x cols.):	Memory	Memory	3,200xMemory
System Features:			
Linear GP	Yes	Yes	Yes
Integer GP	No	Yes	Yes
Nonlinear GP	No	Yes	Yes
Duality	Yes	No	Yes
Sens. Analy.	Yes	Yes	Yes
Other			Fuzzy GP

Table 3-12. (Continued)

Software:	Optimal Engineer	SAS/OR	Solvers
Publisher:	Transpower Corp. 1 Oak Drive Parkerford, PA 19457	SAS Institute Inc. SAS Campus Dr. Cary, NC 27513	Frontline Systs. Inc. P. O. Box 4288 Incline Village, NV 89450
Phone No.:	(800)OPT-TODAY	(919)677-8000	(702)831-0300
FAX No.:	None given	(919)677-8123	(702)831-0314
System Platforms:	Micro	Micro, Mainframe	Micro
Size Capacity (rows x cols.):	Memory	Memory	8,000x4,000
System Features:			
Linear GP	Yes	Yes	Yes
Integer GP	No	Yes	Yes
Nonlinear GP	Yes	No	Yes
Duality	No	Yes	Yes
Sens. Analy.	No	Yes	Yes
Other			

Table 3-12. (Continued)

Software:	SOPT	STORM	XPRESS-MP
Publisher:	Saitech Inc. 1301 Hwy. 36 Hazlet, NJ 07730	Storm Soft. Inc. 24100 Chagrin Blvd. Cleveland, OH 44122-5535	Resource Optim., Inc. 531 S. Gay Str. Ste 1212 Knoxville, TN 37010-1520
Phone No.:	(908)264-4700	(216)464-1209	(615)522-2211
FAX No.:	(908)888-1704	(216)464-4222	(615)522-7907
System Platforms:	Mainframe	Micro	Micro, Mainframe
Size Capacity (rows x cols.):	Memory	600x50	Memory
System Features:			
Linear GP	Yes	Yes	Yes
Integer GP	Yes	Yes	Yes
Nonlinear GP	No	No	Yes
Duality	Yes	No	Yes
Sens. Analy.	No	Yes	Yes
Other			

SUMMARY

This chapter discussed a variety of GP methodology. Included were algorithms and methodology designed to obtain a basic or primary solution for a problem. These primary types of methods included linear GP, integer GP and nonlinear GP. Each of these types of methodologies were subdivided into various other existing methodologies. This chapter also discussed secondary GP methodologies including duality and sensitivity analysis. This chapter closes with a survey of existing software applications that support GP problems.

Like the enumeration techniques of zero-one GP, this chapter seeks to provide researchers with a basic idea of the uniqueness of each method and a listing of available research where in to obtain a comprehensive understanding of the methodologies. The tables in this chapter provide a quick reference guide that is hoped will facilitate research efforts to local relevant information and extend the GP literature.

REFERENCES

All references in this chapter, except those below, can be found in *Appendix B, Journal Research Publications on Goal Programming.*

Beale, E., "Numerical Methods," in Abadie, J., ed., *Nonlinear Programming*, North Holland, Amsterdam, 1967.

Benson, R. G., "*Interactive Multiple Criteria Optimization Using Satisfactory Goals*," Unpublished Ph.D. dissertation, The University of Iowa, 1975.

Bitran, G. R. and Novaes, A. G., "Linear Programming with a Fractional Objective Function," *Operations Research*, Vol. 21 (1973), pp. 22-29.

Box, J., "A New Method of Constrained Optimization and a Comparison with Other Methods," *Computer Journal*, Vol. 8, (1965), pp. 42-52.

Brooke, A., Kendrick, D. and Meeraus, A., *GAMS: A User's Guide*, The Scientific Press, Redwood, CA, 1988.

Charnes, A. and Cooper, W. W., *Management Models and Industrial Applications of Linear Programming*, Vols. 1 & 2, John Wiley and Sons, New York, NY, 1961.

Dantzig G. B. and Wolfe, P., "Decomposition Principle for Linear Programs," *Operations Research*, Vol. 8, No. 1 (1960), pp. 101-111.

Griffith, R. and Stewart, R, "A Nonlinear Programming Technique for the Optimization of Continuous Processing Systems," *Management Science*, Vol. 7 (1961), pp. 370-392.

Fang, S. and Putherpura, S., *Linear Optimization and Extension*, Prentice Hall, Englewood Cliffs, NJ, 1993.

Hwang, C. L. and Masud, A. S. M., eds., *Multiple Objective Decision Making-Methods and Applications*, Springer, Berlin-Heidelberg, 1979.

Ignizio, J. P., *Goal Programming and Extensions*, Heath (Lexington Books), Lexington, MA, 1976.

Lee, S. M., *AB:QM software package*, Allyn and Bacon, Boston, MA, 1993.

Liebman, J., Lasdon, L., Schrage, L. and Waren, A., *Modeling and Optimization with GINO*, The Scientific Press, Palo Alto, CA, 1986.

Martos, B., "Hyperbolic Programming," *Naval Research Logistics Quarterly*, Vol. 11 (1964), pp. 135-155.

Miller, C., "The Simplex Method for Local Separable Programming," in Graves, R. L. and Wolfe, P., eds, *Recent Advances in Mathematical Programming*, McGraw-Hill, New York, NY, 1963.

Saber, H., "*A Partitioning Gradient Based (PGB) Algorithm for Solving Nonlinear Goal Programming Problems*," Unpublished Ph.D. dissertation, The University of Oklahoma, 1991.

Saltzman, M. J., "Broad Selection of Software Packages Available," *OR/MS Today*, Vol. 21, No. 2 (April 1994), pp. 42-51.

Sharda, R., "Linear Programming Software for Personal Computers: 1992 Survey," *OR/MS Today*, Vol. 19, No. 3 (1992), pp. 44-60.

Spronk, J., *Interactive Multiple Goal Programming: Applications to Financial Planning*, Martinus Nijhoff, Boston, MA, 1981.

Turban, E. and Meredith, J. R., *Fundamentals of Management Science*, 5th ed., Irwin, Homewood, IL, 1991

Wynne, A., "*Multicriteria Optimixation with Separable and Chance Constrained Goal Programming*," Unpublished Ph.D. dissertation, University of Nebraska-Lincoln, 1978.

Zoutendijk, G., *Methods of Feasible Directions*, Elsevier, Amsterdam, 1960.

CHAPTER 4. GOAL PROGRAMMING APPLICATIONS

Goal programming (GP) is a very applied methodology. In the over 980 journal citations listed in *Appendix B*, 666 or over 68 percent are applications, case studies or applied models. Indeed, the diversity of application in GP is now so great that just their listing will absorb this entire chapter.

The purpose of this chapter is to help researchers to identify relevant applied GP literature. To accomplish this objective, the chapter will offer users a categorized listing of the GP journal application research. This listing will be organized to permit users an efficient means to identify GP research by type of application.

INTRODUCTION

GP has been used in a wide range of unusually diverse applications, including Christmas tree optimization (Hansen 1978), the pricing of alcoholic beverages (Korhonen and Soismaa 1988), the optimization of fertilizer use (Minguez, Romero and Domingo 1988), and even the rationing of pregnancy (Neal, France and Treacher 1986). GP has also tackled socially serious issues, like busing children for racial equality (Knutson, Marquis and Ricchiute 1980, Lee and Moore 1977, Saunders 1981), the issue of pollution (Nanda, Kothari and Lingamurthy 1988), the social responsibility of business (Kahalas and Satterwhite 1978), and even the philosophical issue of weighting profit vs. social values (Kahalas and Groves 1978).

The magnitude of available applied GP research can make the effort to find relevant research on a particular topic a challenge. Available bibliographies organized by type of GP application are often limited to several dozen categories in which to identify topical areas of application (see Lin 1979, 1980a, Soyibo 1985, Zanakis and Gupta 1985, Romero 1986, 1991, pp. 106-120, White 1990). Also, many of the improvements in methodology appear with or because of their application to real world problems. For many researchers who are interested in research that applies a particular GP methodology (e.g., integer GP, nonlinear GP, etc.), it is particularly time consuming to review the prior literature. Survey contributions by Zanakis and

Gupta (1985) and Romero (1986, 1991, pp. 106-120) have substantially reduced the effort, but are based on a very limited sample of the true population of all GP applied research.

This chapter provides a bibliography of GP application research that is far more substantial than any bibliography of GP applications to date. The studies listed in the tables in this chapter represent actual real world applications or case studies of GP models, as well as applied hypothetical problem applications of GP models.

The listing of the citations in this bibliography are first organized into nine *functional categories* of *accounting, agriculture, economics, engineering, finance, government, international, management* and *marketing*. Each of these nine categories are divided into more than 160 *topical areas* to provide greater detail on the nature of the application. To permit users to identify the type of GP model that is used in each article, a coding system is utilized. Citations that have one asterisk "*" utilize an integer GP model (i.e.., all integer GP, mixed integer GP or zero-one GP). Citations that have two asterisks "**" utilize some type of nonlinear GP model. Citations that have three asterisks "***" utilize one or more of the other methodologies discussed in Chapter 3 (i.e., fuzzy GP, chance constraint GP, etc.). Citations without an asterisk generally utilize a weighted, preemptive or combination type GP model.

To insure that users will be able to find all existing relevant citations under a specific topic, citations in this bibliography have been cross-listed. This cross-listing permits the same article to appear under multiple topic headings and in multiple categories when appropriate.

GOAL PROGRAMMING APPLICATIONS IN ACCOUNTING

The 38 total GP types of applications in accounting are organized into ten different topics. The citations are presented in Table 4-1. The *Other Accounting* topic is used here for citations that could not easily be identified and classified into the other topical areas. The remaining nine topic areas are self explanatory.

Table 4-1. Citations on GP Applications in Accounting

Topic	Available References on Topic
Assets	Hibiki and Fukukawa (1992), Philippatos and Christofi (1984), Rosenbloom and Shiu (1990), Siokas and Vassiloglou (1991)
Auditing	Bailey, Boe and Schnack (1974), Balachandran and Steuer (1982)***, Blocher (1978), El-Sheshai, Harwood and Hermanson (1977)*, Filios (1984), Gardner, Huefner and Lotfi (1990), Ijiri and Kaplan (1971), Tayi and Gangolly (1985)***
Balance Sheet	Eatman and Sealey (1979), Tayi and Leonard (1988)
Budgeting	Buffa (1983)*, Charnes and Stedry (1966), Charnes, Colantoni, Cooper and Kortanek (1972), Charnes, Cooper and Ijiri (1963), Chen (1983)*, Joiner and Drake (1983), Kwak and Diminnie (1987), Lin (1978), Lootsma, Mensch and Vos (1990), Olve (1981), Smith (1978)
Control Systems	Lin (1980b)
Cost	Badran (1984), Charnes, Colantoni and Cooper (1976), Pentzaropoulos and Gilkas (1993), Sheshai, Harwood and Harmanison (1977)*
Other Accounting	Kornbluth (1974), Lin (1979), Welling (1977)
Public Accounting	Garrod (1991), Killough and Souders (1973)
Taxes	Lee, Lerro and McGinnis (1971), Puelz and Lee (1992)
Transfer Pricing	Merville and Tavis (1974)

*Integer GP model
**Nonlinear GP model
***Other specialized GP model

GOAL PROGRAMMING APPLICATIONS IN AGRICULTURE

The 63 total GP types of applications in agriculture are organized into ten different topics. The citations are presented in Table 4-2. The *Other Agriculture* topic is used here for citations that could not easily be identified and classified into the other topical areas. The remaining nine topic areas are self explanatory.

Table 4-2. Citations on GP Applications in Agriculture

Topic	Available References on Topic
Aquaculture	Weithman and Ebert (1981), Shepherd (1981), Sandiford (1986), Muthukude, Novak and Jolly (1990), Drynan and Sandiford (1985), Everitt, Sonntag, Puterman and Whalen (1978)
Economics	Bazaraa and Bouzaher (1981), Bouzaher and Mendoza (1987)
Farming	Barnett, Blake and McCarl (1982), Berlo (1993), Dobbins and Mapp (1982)***, Drynan (1985), Eto (1991), Fahmy and El-Shishiny (1991), Flinn, Jayasuriya and Knight (1980), Kang (1983), McCarl and Blake (1983), Minguez, Romero and Domingo (1988), Patrick and Blake (1980), Piech and Rehman (1993), Rehman and Romero (1993), Romero and Rehman (1983, 1984a, 1985), Varshney and Rao (1989)

*Integer GP model
**Nonlinear GP model
***Other specialized GP model

Table 4-2. (Continued)

Topic	Available References on Topic
Forestry	Arp and Lavigne (1982), Buongiorno and Svanqvist (1982)***, Chang and Buongiorno (1981), Dyer, Hof, Kelly, Crim and Alward (1979, 1983), Field (1973), Field, Dress and Fortson (1980), Hotvedt (1983), Hotvedt, Leushner and Buhyoff (1982), Hrubes and Rensi (1981), Kao and Brodie (1979), McKillop and Liu (1990), Mendoza, Bare and Campbell (1987), Mitchell and Bare (1981)***, Pickens and Hof (1991)***, Porterfield (1976), Rensi and Hrubes (1983), Schuler and Meadows (1975), Schuler, Webster and Meadows (1977), Cubbage, Field, Eza and Farkas (1987), Dane, Meador and White (1977)
Land Management	El-Shishiny (1988), Ghosh, Paul, and Basu (1993)
Other Agriculture	Romero and Rehman (1983), Sinha, Rao and Mangaraj (1988)***, Wheeler and Russell (1977)
Pest Control	Brown, McClendon and Akbay (1990), Johnson, Oltenacu, Kaiser and Blake (1991)
Ranching	Barlett and Clawson (1978), Lara and Romero (1992)***, Neal, France and Treacher (1986), Rehman and Romero (1984, 1987)
Regional Planning	De Wit, Van Keulen, Seligman and Spharim (1988)**
Storage	Chang, Chung and Hwang (1983, 1984)**

*Integer GP model
**Nonlinear GP model
***Other specialized GP model

GOAL PROGRAMMING APPLICATIONS IN ECONOMICS

The 28 total GP types of applications in economics are organized into nine different topics. The citations are presented in Table 4-3. The *Other Economics* topic is used here for citations that could not easily be identified and classified into the other topical areas. The remaining eight topic areas are self explanatory.

Table 4-3. Citations on GP Applications in Economics

Topic	Available References on Topic
Exporting	Levary (1986a)
Income Redistribution	Charnes, Duffuaa and Intriligator (1984)***
Industrial Development	Lee, Tang, Olson and Yen (1989)*, Walker and Chandler (1978)
Municipal Planning	Lee and Sevebeck (1971)
Other Economics	Dynan and Sandiford (1985), Kao and Brodie (1979), Lonergan and Cocklin (1988), Schinnar (1976), Shim (1983)**, Spivey and Tamura (1970), Spronk and Veeneklaas (1983)***, Charnes, Colantoni, Cooper and Kortanek (1972), Samouilidis and Pappas (1980), Charnes, Duffuaa and Al-Saffar (1989)
National Policies	Aggarwal and Clark (1978), Budavei (1982), Charnes and Collomb (1972)***, Habeeb (1991), Kalu (1994), Walleniius (1982), Charnes, Cooper, Harrald, Karwin and Wallace (1976)***
Pollution	Nanda, Kothari and Lingamurthy (1988)

*Integer GP model
**Nonlinear GP model
***Other specialized GP model

Table 4-3. (Continued)

Topic	Available References on Topic
Regional Planning	Chen (1986), Lee and Olson (1981), Tyagi and Swarup (1979), Wright, Revelle and Cohon (1983)
Resource Allocation	Charnes, Cooper, Harrald, Karwan and Wallace (1976)

GOAL PROGRAMMING APPLICATIONS IN ENGINEERING

The 25 total GP types of applications in engineering are organized into eight different topics. The citations are presented in Table 4-4. The *Other Engineering* topic is used here for citations that could not easily be identified and classified into the other topical areas. The remaining seven topic areas are self explanatory.

Table 4-4. Citations on GP Applications in Engineering

Topic	Available References on Topic
Automated Systems	Bard (1986)
Design Problems	Bascaran, Mistree and Bannerot (1987), Guven, Mistree and Bannerot (1984), Ignizio and Satterfield (1977), Ignizio (1981a), Ignizio (1987), Ishiyama, Hondoh, Ishida and Onuki (1989), Kornbluth (1986)**, McCammon and Thompson (1980, 1983), Singh and Agarwal (1983), Singh (1983)
Feasibility Study	Ng (1991)

**Nonlinear GP model

Table 4-4. (Continued)

Topic	Available References on Topic
Other Engineering	Qassim and Silveira (1988), Van Crombrugge and Thompson (1985), Wang (1986), Siha (1993)
Production Processing	Acharya, Jain and Batra (1986)
Routing	Elamin, Duffuaa and Yassein (1990)
Reliability	Gen, Tsujimura and Chang (1993)***, Gen, Ida and Lee (1989, 1990)*, Hwang, Lee, Tillman and Lie (1984)**, Jedrzejowics and Rosicka (1983)
Software Application	Svestka (1990)

*Integer GP model
**Nonlinear GP model
***Other specialized GP model

GOAL PROGRAMMING APPLICATIONS IN FINANCE

The 112 total GP types of applications in finance are organized into 17 different topics. The topic areas are self explanatory. The citations are presented in Table 4-5.

Table 4-5. Citations on GP Applications in Finance

Topic	Available References on Topic
Acquisition Analysis	Hoffman and Schniederjans (1990), Madey and Dean (1985), Schniederjans and Fowler (1989), Schniederjans and Hoffman (1992)*, Dean and Schniederjans (1991), Hoffman, Schniederjans and Sirmans (1990)
Banking	Bandyopadhyay (1978), Fortson and Dince (1977), Giokas and Vassiloglou (1991), Hibiki and Fukukawa (1992), Hollis and Murray (1985), Keown (1978)***, Siokas and Vassiloglou (1991), Wilstead, Hendrick and Stewart (1975), Zaloom, Tolga and Chu (1986), Zanakis, Mavrides and Roussakis (1986)
Bank Portfolios	Booth and Dash (1977), Booth and Dash (1977)**, Chambers and Charnes (1961), Korhonen (1987)
Bond Portfolios	Alexander and Resnick (1985), Lee and Puelz (1989), Sealey (1977)
Capital Budgeting	Chateau (1975), Choi and Levary (1989)***, De, Acharya and Sahu (1982, 1986)***, Deckro, Spahr and Hebert (1985), Forsyth (1969), Hawkins and Adams (1974), Ignizio (1976), Jackman (1973), Keown and Martin (1977)***, Keown and Martin (1976, 1978)*, Keown and Taylor (1980)***, Keown and Taylor (1981), Lawrence and Reeves (1982b), Lee and Lerro (1974b), Lee and Shim (1984), Taylor and Keown (1981)*, Thanassoulis (1985)

*Integer GP model
**Nonlinear GP model
***Other specialized GP model

Table 4-5. (Continued)

Topic	Available References on Topic
Capital Flow	Arthur and Lawrence (1985), Sartoris and Spruill (1974), Lee and Lerro (1974a)
Credit Analysis	Srinivasan and Kim (1987)
Divestiture	Sueyoshi (1991)***, Charnes, Cooper and Sueyoshi (1988), Evans and Heckman (1988)
Financial Planning	Ashton and Atkins (1979),Callahan (1973),Cook (1984),Jaaskelainen and Lee (1971), Kvanli and Buckley (1986), Kvanli (1980), Lee and Eom (1984, 1989), Lee, Justis and Franz (1979), Linke and Whitford (1983), Merville and Tavis (1974), Miyajima and Nakai (1986)***, Puelz and Lee (1992), Sealey (1978), Vinso (1982)
Finance/Production	Forsyth (1969)
Global Financial Planning	Eom and Lee (1987), Eom, Lee and Snyder (1987), Hoffman and Schniederjans (1990), Merville and Petty (1978), Philippatos and Christofi (1984), Schniederjans and Hoffman (1992)*, Vinso (1982)
Insurance	Drandell (1977), Flock and Lee (1974), Lawrence and Reeves (1982b), Lee and Klock (1974), Lilly and Gleason (1976)
Investment planning	Caplin and Kornbluth (1975), Chae, Suver and Chou (1985), Chen (1987)***, Gleason and Lilly (1977), Hsu (1976), Jose and Tabucanon (1986), Kumar and Philippatos (1979), Leinbach and Cromley (1983), Wacht and Whitford (1976)

*Integer GP model
**Nonlinear GP model
***Other specialized GP model

Table 4-5. (Continued)

Topic	Available References on Topic
Managing Risk	Booth and Bessler (1989), Hibiki and Fukukawa (1992), Hollis and Murray (1985), Lee and Hall (1988), Sharda and Musser (1986)
Mutual Funds Portfolio	Lee and Lerro (1973)
Other Finance	Ashton (1985, 1986), Batson (1989), Hong (1981), Jones (1979), Kornbluth and Vinso (1982)***, Orne and Rao (1975), Walker and Chandler (1978), Joiner (1981)
Portifolio Analysis	Harrington and Fischer (1980)***, Johnson, Zorn and Schniederjans (1989), Kumar and Philippatos (1978), Lee and Chesser (1980), Levary and Avery (1984), Muhlemann and Lockett (1980)***, Muhlemann, Lockett and Gear (1978)***, Schniederjans, Zorn and Johnson (1993), Stone and Reback (1975)

*Integer GP model
**Nonlinear GP model
***Other specialized GP model

GOAL PROGRAMMING APPLICATIONS IN GOVERNMENT

The 169 total GP types of applications in government are organized into 27 different topics. The citations are presented in Table 4-6. The broadness of this category required an additional level of categorization. The citations are first divided into three broad categories of government involvement: *Education, Health Care* and *General Government*. The *Education-General* and *Health Care-General* topics are used here for citations that could not easily be identified and classified into the other topical areas. The remaining 25 topic area headings are self explanatory.

Table 4-6. Citations on GP Applications in Government

Topic	Available References on Topic
EDUCATION:	
Education-General	Djang (1993), Joiner (1980), Kennedy (1991), Smith (1978), Thomas (1987), Walters, Mangold and Haran (1976)
Eduation-Library	Beilby and Mott (1983), Hannan (1978a), Schniederjans and Santhanam (1989)*
Education-Secondary	Sutcliffe and Board (1986), Sutcliffe, Board and Cheshire (1984), Cohn and Morgan (1978), Papageorgia (1978)
Education-Students	Bafail (1993), Campbell and Ignizio (1972), Miyaji, Ohno and Mine (1988), Rumpf (1987)
Education-University	Albright (1975), Benjamin, Ehir and Omurtag (1992), Chae, Suver and Chou (1985), Diminnie and Kwak (1986), Feuer (1985), Franz, Lee and Van Horn (1981), Geoffrion, Dyer and Fienberg (1972)***, Ghosh, Paul and Basu (1992), Greenwood and Moore (1987), Harwood and Lawless (1975), Kendall and Luebbe (1981), Keown, Taylor and Pickerton (1981), Kwak and Diminnie (1987), Lawrence, Lawrence and Reeves (1983), Lee and Clayton (1970a,1972, 1980a, 1980b), Lee and Moore (1974b, 1975), Lee and Soyibo (1986), Lee and Van Horn (1978), McClure and Wells (1978b), Min (1988b)***, Ritzman and Krajewski (1979), Schniederjans and Kim (1987)*, Schroder (1974), Soyibo and Lee (1986)

*Integer GP model
**Nonlinear GP model
***Other specialized GP model

Table 4-6. (Continued)

Topic	Available References on Topic
HEALTH CARE:	
Health Care-Blood	Kendall and Lee (1980a, 1980b), Kendall (1980)
Health Care-Budgeting	Keown and Martin (1976)*, Trivedi (1981)*, Mersha, Meredith and McKinney (1987)
Health Care-Dieting	Anderson and Earle (1983), McCann-Rugg, White and Endres (1983), Romero and Rehman (1984b)
Heath Care-General	Christainson (1983), Lee and Lerro (1974a), McGlone and Calantone (1992), Nelson and Wolch (1985), Ozatalay and Broyles (1984), Parker (1983), Rifai and Pecenka (1990), Tingley and Liebman (1984a), Wilson and Gibberd (1990)
Health Care-Hospitals	Chae, Suver and Chou (1985), Ghandfaroush (1993), Lee (1973), Wacht and Whitford (1976), Butler, Karwan, Sweigart and Reeves (1992)
Health Care-Nurse	Arthur and Ravindran (1981), Chen and Yeung (1993), Ozkarahan Bailey (1988), Ozkarahan (1989), Trivedi (1981)*, Moores, Garrod and Briggs (1978), Musa and Saxena (1984)
Health Care-System Design	Charnes and Storbeck (1980), Franz, Baker, Leong and Rakes (1989), Franz (1989), Franz, Rakes and Wynne (1984)***, Parker, Mtango, Koda and Killewo (1986), Specht (1993), Zhu and McKnew (1993), Baker and Fitzpatrick (1986), Baker, Clayton and Taylor (1989)**, Baker, Clayton and Moore (1989)

*Integer GP model
**Nonlinear GP model
***Other specialized GP model

Table 4-6. (Continued)

Topic	Available References on Topic
GENERAL GOVERNMENT:	
Allocating Resources	Buffa and Shearon (1980), Koizumi and Inakazu (1989), Simkin (1977), Cornett and Williams (1991), Donckels (1977)**, Dusansky and Kalman (1981)**, Sinha and Sastry (1987b), Sinha, Sastry and Misra (1988)
Environmental Issues	Cocklin, Lonergan and Smit (1986), Charnes, Cooper, Karwan and Wallace (1979)***, Charnes, Cooper, Harrald, Karwan and Wallace (1976)***, Charnes, Haynes, Hazleton and Ryan (1975), Cocks and Baird (1989), Harrald, Wallace and Wendell (1978), Kambo, Handa and Bose (1991), Lonergan and Cocklin (1988), Ludwin and Chamberlain (1989), Marten and Sancholuz (1982), Panagiotakopoulos (1975), Spronk and Veeneklaas (1983), Werczberger (1976)
Government Budgeting	Joiner and Drake (1983)
Military	Armstrong and Cook (1979), Bres, Burns, Charnes and Cooper (1980), Gallagher and Kelly (1991), Lee, Synder and Brisch (1983), Mellichamp, Dixon and Mitchell (1980), Morey (1985), Nussbaum (1980), Suzuki and Yoshizawa (1994)***, Taylor, Keown and Greenwood (1983)*, Weigel and Wilcox (1993), Collons, Gass and Rosendahl (1983), Gass (1991), Gass and Collins (1988)
Police Allocation	Saladin (1982), Taylor, Moore, Clayton, Davis and Rakes (1985)**, Lee, Franz and Wynne (1979)

*Integer GP model
**Nonlinear GP model
***Other specialized GP model

Table 4-6. (Continued)

Topic	Available References on Topic
GENERAL GOVERNMENT (Continued):	
Policy Evaluation	Chicoine, Scott and Jones (1980), Smith (1980), Piekanen (1970)
Policy Compliance	Taylor, Davis and Ryan (1977)
Postal Service Scheduling	Ritzman and Krajewski (1973)
Prison Management	Dessent and Hume (1990)
Social Issues	Kahalas and Groves (1978), Kahalas and Key (1974), Kahalas and Satterwhite (1978), Knutson, Marquis and Ricchiute (1980), Lee and Moore (1977), Nanda and Lingamurthy (1988), Saunders (1981), Zanakis (1987)*
Transportation	Prakash, Sinha and Sahay (1984), Wilson and Gonzalez (1985)
Urban Planning	Courtney, Klastorian and Reufli (1972)m Ignizio (1980), Kambo, Handa and Bose (1991), Lee and Keown (1979), Stern (1974), Taylor and Keown (1978b), Taylor (1977)
Utility Management	Chetty Mallikarjuna and Subramanian (1988), Linke and Whitford (1983)
Waste Management	Sushil and Vrat (1989), Sushil and Agrawal (1989), Sushil (1993), Mogharabi and Ravindran (1992)

*Integer GP model
**Nonlinear GP model
***Other specialized GP model

Table 4-6. (Continued)

Topic	Available References on Topic
GENERAL GOVERNMENT (Continued):	
Water Resource Mgt.	Can and Houck (1984), Chisman and Rippy (1977), Changchit and Terrell (1993), Dauer and Kruger (1980), Giocoechea, Duckstein and Fogel (1976), Houck (1985), Jain, Soni and Seethapathi (1988), Loganathan and Bhattacharya (1990), Lohani and Adulbhan (1979), Loucks (1977), McGregor and Dent (1993), Monarchi, Kisiel and Duckstein (1973)***, Neely, North and Fortson (1976,1977), Neely, Sellers and North (1980), Nelson (1979), Reznicek, Simonovic and Bector (1991), Szidarovsky and Duckstein (1986), Taylor, Davis and North (1975), Yazadanian and Peralta (1986)

*Integer GP model
**Nonlinear GP model
***Other specialized GP model

GOAL PROGRAMMING APPLICATIONS IN AN INTERNATIONAL CONTEXT

The 42 total GP types of applications in an international context are organized into seven different topics. The citations are presented in Table 4-7. The topic areas are structured in terms of the major categories used in this book. Since these citations are cross-listed, their listing here allows users an additional international dimension to sort on for identification purposes.

Table 4-7. Citations on GP Applications an International Context

Topic	Available References on Topic
Accounting	Merville and Petty (1978), Philippatos and Christofi (1984)
Agriculture	Barnett, Blake and McCarl (1982), Bazaraa and Bouzaher (1981), Buongiorno and Svanqvist (1982), Chang, Chung and Hwang (1983)**, Chang, Chung and Hwang (1984)**, Charnes, Duffuaa and Al-Saffar (1989), Kang (1983), Lara and Romero (1992)***, Romero and Rehman (1983)
Economics	Habeeb (1991), Kalu (1994)
Engineering	Guven, Nistree and Bannerot (1984)
Finance	Bandyopadhyay (1978), Eom and Lee (1987), Hoffman and Schniederjans (1990), Lee and Eom (1989), Lee and Eom (1984), Leinbach and Cromley (1983), Levary and Avery (1984), Merville and Petty (1978), Philippatos and Christofi (1984), Schniederjans and Hoffman (1992)*, Vinso (1982), Hollis and Murry (1985), Hollis (1978), Kornbluth and Vinso (1982)***
Government	Anderson and Earle (1983),Chae, Newbrander and Thomason (1989), Cocks and Baird (1989), Everitt, Sonntag, Puterman and Whalen (1978), Jain, Soni and Seethapathi (1988), Marten and Sancholuz (1982), McGregor and Dent (1993), Nelson (1979), Papageorgio (1978), Parker, Mtango, Kado and Killewo (1986)
Management	Al-Faraj, Alidi and Al-Ibrahim (1993),Chang, Chung and Hwang (1983), Chang, Chung and Hwang (1984), Ehir and Benjamin (1993)

*Integer GP model
**Nonlinear GP model
***Other specialized GP model

GOAL PROGRAMMING APPLICATIONS IN MANAGEMENT

The 244 total GP types of applications in management are organized into 45 different topics. The citations are presented in Table 4-8. The broadness of this category required an additional level of categorization. The citations are first divided into three broad categories of government involvement: *Human Resource Management, Management Information Systems* and *Production and Operations Management*. The *Other Human Resources* and *Other MIS* topics are used here for citations that could not easily be identified and classified into the other topical areas. The remaining topic areas headings are self explanatory.

Table 4-8. Citations on GP Applications in Management

Topic	Available References on Topic
HUMAN RESOURCE MANAGEMENT:	
Assigning and Scheduling	Charnes, Cooper, Niehaus and Stedry (1969), Jones and Kwak (1982), Koelling and Bailey (1984), Lawrence, Reeves and Lawrence (1984), Loucks and Jacobs (1991), Mehta and Rifai (1976), Min (1987b)*, Zanakis and Lawrence (1977)*, Zanakis and Maret (1981a), Zanakis (1983)***, Organization structure Freeland (1976), Lee and Rho (1985)***, Lee, Luthans and Olson (1982)
Compensation	Charnes, Cooper and Ferguson (1955), Kwak, Allen and Schniederjans (1982), Steuer (1983)
Corporate Evaluation	Sridhar and Raghavendra (1988)

*Integer GP model
**Nonlinear GP model
***Other specialized GP model

Table 4-8. (Continued)

Topic	Available References on Topic
HUMAN RESOURCE MANAGEMENT(Continued):	
Human Resource Planning	Aronson and Thomson (1985), Charnes, Cooper and Niehaus (1975),Henderson (1982),Kornbluth (1983), Martel and Price (1981)**, Min (1990), Price and Gravel (1984), Price and Piskor (1972), Price (1978, 1980), Taylor, Moore and Clayton (1982)**, Whitford and Davis (1983), Wijngaard (1987)
Job Evaluation	Gupta and Ahmed (1988)
Labor Negociation	Schniederjans and Kim (1987)
Organizational Decisions	Bonczek, Holsapple and Whinston (1979), Keown and Taylor (1978)*, Kimory (1978), Lee and Litschert (1976), Lee and Shim (1986)***, Sweeney, Winkofsky, Roy and Baker (1978), Tersine (1976)
Other Human Resources	Bottoms and Bartlett (1975), Kananen, Korhonen, Wallenius and Wallenius (1990), Parker and Kaluzny (1982), Rifai (1978)
Policy Evalution	King (1982), Rivett (1977)
Productivity Measurement	Sardana and Vrat (1987)
R&D Allocations	Keown, Taylor and Duncan (1979)*, Khorramshahgol and Gousty (1986), Khorramshahgol, Dabzie and Akaah (1988), Kwak and Jones (1978), Ringuest and Grayes (1989), Stewart (1991)

*Integer GP model
**Nonlinear GP model
***Other specialized GP model

Table 4-8. (Continued)

Topic	Available References on Topic
Strategic Planning	Jaaskelainen (1972)
MANAGEMENT INFORMATION SYSTEMS:	
Decision Support Systems	Khorramshahgol and Azani (1988)
Design	Henderson and Schilling (1985), Ignizio, Plamer and Murphy (1982), Jain and Dutta (1986), Jain (1984), Lee and Wilkins (1983)
Distributed Data	Chen, Farn and Tsay (1991)*, Ling-Hwie, Kwo-Jean and Ching-Shu (1991)*
Evalution Systems	Chandler (1982), Djang (1993), Parker (1985), Tanner (1991)
Microcomputers	Lee and Shim (1984)
Other MIS	Henderson and Schilling (1985), Gross and Talavage (1979)
Project Selection	Santhanam, Muralidhar and Schniederjans (1989), Schniederjans and Santhanam (1993a, 1993b)*, Schniederjans and Wilson (1991)*
Time Sharing	Dyer (1973)
PRODUCTION AND OPERATIONS MANAGEMENT:	
Acquisition Planning	Utar and Schoenfled (1973)

*Integer GP model
**Nonlinear GP model
***Other specialized GP model

Table 4-8. (Continued)

Topic	Available References on Topic
PRODUCTION AND OPERATIONS MANAGEMENT (Continued):	
Aggregate Planning	Deckro and Hebert (1988)**, Gilgeous (1989), Goodman (1974), Hindelang and Hill (1978), Jaaskelainen (1969), Lockett and Muhlemann (1978), Musud and Hwang (1980), Rakes, Franz and Wynne (1984), Welman (1976)
Assembly Line	Baybars (1985), Fisher, Wei and Dontamsetti (1989), Gunther, Johnson and Peterson (1983), Malakooti (1991b), Decko and Rangachari (1990)
Blending Problems	Arthur and Lawrence (1980), Lee and Olson (1983)***
Design	Osinski, Pokojski and Wrobel (1983), Sankaran (1990), Shafter and Rogers (1991), Shih and Hagels (1989)*, Singh and Verma (1985), Singh, Aneja and Rana (1990), Eilon (1982), Fortenberry, Mitra and Willis (1989), Green, Kim and Lee (1981)
Energy	Bit and Alam (1993), Choobineh and Burgman (1984)
Flexible Manufacturing	Chen and Askin (1990), Dean, Schniederjans and Yu (1990), Frazier, Gaither and Olson (1990), Gangan, Khator and Bahu (1987), Imany and Schlesinger (1989), Kumer, Singh and Tewari (1987, 1991)***, Lee and Jung (1989), O'Grady and Mennon (1984, 1986a, 1986b), Ro and Kim (1990)

*Integer GP model
**Nonlinear GP model
***Other specialized GP model

Table 4-8. (Continued)

Topic	Available References on Topic
PRODUCTION AND OPERATIONS MANAGEMENT (Continued):	
Forecasting	Hattenschwiler (1988), Reeves and Lawrence (1982)
Inventory and Distribution	Brauer and Naadimuth (1992), Buffa (1976, 1983), Chaudhry, Forst and Zydiak (1991), El-Dash (1992)**, Golany, Yadin and Learner (1991), Karmarkar (1979)**, Mehrez and Ben-Arich (1991)***, Padmanabhan and Vrat (1990)**, Rao (1980), Sinha and Sastry (1987a), Wascher (1990)
Kanban/Just-In-Time	Fukukawa and Hong (1993), Kim and Schniederjans (1993), Lee, Chung and Everett (1992)
Loading and Handling	Ng (1992), Osleeb and Ratick (1983)
Location and Layout	Al-Faraj, Alidi and AL-Ibrahim (1993), Alonso and Devaux (1981), Barda, Dupuis and Lecioni (1990), Benito and Devaux (1981), Bhattachary, Rao and Tiwari (1992,1993)***, Current, Min and Schilling (1990), Davis, Stam and Grzybowski (1992), Dieperink and Nijkamp (1987), Kwak and Schniederjans (1985a, 1985b), Lee and Franz (1979), Lee and Lubbe (1987a), Lee and Schniederjans (1983), Lee, Green and Kim (1981), Min (1988a), Mohanty and Rathnakumar (1984), Nijkamp and Spronk (1979), Rosenblatt (1979), Schniederjans, Kwak and Helmer (1983), Venugopal and Mothanty (1982), Solomon and Haynes (1984), Solanki (1991)

*Integer GP model
**Nonlinear GP model
***Other specialized GP model

Table 4-8. (Continued)

Topic	Available References on Topic
PRODUCTION AND OPERATIONS MANAGEMENT (Continued):	
Logistics and Routing	Arthur and Lawrence (1982), Gingrich and Soli (1984), Johnson (1976), Lawrence and Burbridge (1976), Park and Koelling (1986), Srinivasan and Thompson (1972)
Lot-Sizing	McKnew and Sauydan (1991)*, Vickery and Markland (1985, 1986)*
Maintenance	Armstrong and Cook (1981)
Materials Management	Mohanty and Chandran (1984)
Production/Finance	Forsyth (1969)
Production/Marketing	Hansen (1978), Taylor and Anderson (1979)
Production Planning	Chanda (1990), Ehir and Benjamin (1993), Fisk (1979), Golovanov, Zotov, Maikov and Pushnyak (1987), Green, McCarthy and Pearl (1983), Kendall and Schniederjans (1985), Mackulak, Moodie and Williams (1980), Malakooti (1989, 1991a)***, Miller and Davis (1978), Mohanty and Govindrajan (1989), Mohanty and Singh (1992), Philipson and Ravindran (1978), Ray (1986), Ruefli and Storbeck (1984), Salvia (1979), Sarma, Sellami and Houam (1993), Schniederjans and Markland (1986), Sullivan and Fitzsimmons (1978), Sundaram (1978), Tabucanon and Mukyangkoon (1985)***, Utar and Schoenfled (1973), Zanakis and Smith (1980)

*Integer GP model
**Nonlinear GP model
***Other specialized GP model

Table 4-8. (Continued)

Topic	Available References on Topic
PRODUCTION AND OPERATIONS MANAGEMENT (Continued):	
Project Planning	Deckro and Hebert (1984), Hannan (1978b), Moore, Talyor, Clayton and Lee (1978), Rakes and Franz (1985), Taylor and Keown (1978a), Lee, Park and Economides (1978)
Purchasing	Buffa and Jackson (1983)
Quality	Bardaro and Mutmansky (1986), Bishop, Narayanan and Greney (1977), Ebrahimpour and Ansari (1988), Hindelang (1973), Irani, Mittal and Lehtihet (1989), Kalro, Chaturvedi and Sengupta (1983), Klimberg, Revelle and Cohon (1991), Rifai and Dey (1990), Sengupta (1981), Tayi (1985)**, Williams and Zigli (1987)
Sampling	Chakraborty (1986, 1988a, 1988b, 1991a, 1991b)***, Drezner and Wesolowsky (1991), Ravindran, Shin, Arthur and Moskowitz (1986)**
Scheduling Production	Chen (1988)*, Crouch (1984), Daniels (1990), Dary-Dowman and Mitra (1985)***, Deckro, Hebert and Winkofsky (1984), Fisk (1980), Hershauer and Gowens (1977)*, Jacobs and Wright (1988), Lashine, Foote and Ravindran (1991)**, Lee and Clayton (1970b), Lee and Moore (1974a), Lee and Olson (1984)*, Lee, Clayton and Taylor (1978), Selen and Hott (1986)*, Utar and Schoenfled (1973), Van Hulle (1991b)*, Wilson (1989)

*Integer GP model
**Nonlinear GP model
***Other specialized GP model

Table 4-8. (Continued)

Topic	Available References on Topic
PRODUCTION AND OPERATIONS MANAGEMENT (Continued):	
Transportation	Acharya, Nayak and Mohanty (1987), Al-Faraj, Alidi and AL-Ibrahiam (1993), Hemaida and Kwak (1994), Kwak and Schniederjans (1979, 1985d), Lee and Moore (1973), Moore, Taylor, Clayton and Lee (1978), Morse and Clark (1975), Nayak, Basu and Tripathy (1989), Singh and Kishore (1991), Stewart and Ittman (1979), Turshen and Wester (1986)

*Integer GP model
**Nonlinear GP model
***Other specialized GP model

GOAL PROGRAMMING APPLICATIONS IN MARKETING

The 24 total GP types of applications in marketing are organized into eleven different topics. The citations are presented in Table 4-9. The *Other Marketing* topic is used here for citations that could not easily be identified and classified into the other topical areas. The remaining ten topic areas are self explanatory.

Table 4-9. Citations on GP Applications in Marketing

Topic	Available References on Topic
Distribution Channels	Kim (1983), Kwak, Schniederjans and Warkentin (1991)
Marketing/Production	Hansen (1978), Taylor and Anderson (1979)

Table 4-9. (Continued)

Topic	Available References on Topic
Market Segmentation	McGlone and Calantone (1992)
Media Planning	Charnes, Cooper, DeVoe, Learner and Reinecke (1968), Charnes, Cooper, Learner and Snow (1968), Clayton and Moore (1972a), De Kluyver (1978, 1979a), Dyer, Forman and Mustafa (1992), Keown and Duncun (1979)*
Other Marketing	Lee and Nicely (1974), Mehta and Riafai (1979), Wilson (1975)
Pricing	Brown and Norgaard (1992), Korhonen and Soismaa (1988)
Product Development	Bard (1990)
Purchasing	Williams (1987)
Retailing	Kim (1983), Min (1987a)
Sales Management	Lee and Bird (1970), Mahajan and Valharia (1990), McClure and Wells (1987a)
Warranty Estimation	Mitra and Patankar (1988, 1990, 1993)

*Integer GP model

SUMMARY

This chapter lists the GP applied publications that have appeared in the literature since GP was first created in 1955. The listing is this chapter is based on those journal citations that appear in *Appendix B* of this book. The fact that only journal citations are being used is a possible limitation on the comments that follow.

As can be seen in this chapter, a lot of GP applied research has been done. This chapter should also be viewed as a call for a lot more GP research to be done. Of the 666 actual citations listed in *Appendix B*, a total of 745 cross-listed applications appear in the tables in this chapter. If the number of citations observed in these tables represent the true proportions in the population of all GP applied research, then the relative creation of research into the nine major categories is as stated in Table 4-10.

Table 4-10. Citation Counts, Ranking and Proportion of GP Applications

Category	Citation Counts	Ranking	Proportion (%)*
Management	244	1	32.8
Government	169	2	22.7
Finance	112	3	15.0
Agriculture	63	4	8.5
International	42	5	5.6
Accounting	38	6	5.1
Economics	28	7	3.7
Engineering	25	8	3.4
Marketing	24	9	3.2
Total	745		100.0

*Based on proportion of total citation counts.

A point is worth mentioning on the relative proportions of publications in Table 4-10. One way of viewing this table is in its presentation of possible publication opportunities. The opportunities for a GP publication in each category area (i.e., in the respective journals in each area) can possibly be viewed in a type of inverse relationship. That is, journals in the areas of marketing and engineering might whole the greatest opportunity for a publication, over journals in management. This doesn't mean that all management journals should not be considered as possible outlets, only that in general it may be easier to get publications in nonmanagement journals where virgin ideas for applications are still more plentiful.

Since many of the researchers who work with GP are *management science* or *operations research* faculty, it is not surprising to find the largest number of publications is in the *management* area. The fact that one has to publish in one's own departmental areas to become tenured or promoted is also no surprise. Interestingly enough, most college and university accreditation organizations want colleagues from differing departments to work together on research and teaching. They feel that such cross-pollination will help to generate better researchers and teachers. It might also help to generate more GP publications.

CHAPTER 5. FUTURE TRENDS IN GOAL PROGRAMMING

Goal programming (GP) research has had a great past and will have a great future in helping to improve decision making. Trends in the types of GP research that appear in the literature are important to observe in order to take part in making contributions to that future.

The purpose of this chapter is summarize and conclude this book's presentation on GP. This is accomplished with commentary and suggestions on trends in GP research. This chapter will present a commentary on how GP is currently positioned for growth in research. This section will be followed by a presentation on suggestions for future GP research.

GP IS POSITIONED FOR GROWTH

To provide for continual growth, the *life cycle of GP research* must be able to demonstrate an ability to overcome obstacles to grow and maintain the interest of individuals who will contribute to that growth. This section focuses on these two issues to help researchers understand just how well the life cycle of GP research is positioned for a growth stage.

Overcoming Past Obstacles to Growth

Two major obstacles to the GP research growth that are commonly repeatedly in the literature concern a lack of theoretical developments and the lack of computer software. Zanakis and Gupta (1985) claimed that a lack of sufficient new theoretical developments was possibly a major reason for a decline in GP publications in the mid-1980's. The reality was that GP in 1984 when Zanakis and Gupta were researching their paper had not even reached its peak of publications which occurred in 1987. Their call for more theoretical and empirical work on GP was substantially answered by dozens of studies on priority preferences, goal trade-off analyses and weightings methodologies (see Tables 2-3 and 2-4). The result on total GP research publications was a substantial and sustained increase of GP publications running through the year 1991 (see Table 1-3). The theoretical contributions currently available, but not always used, is more than adequate to deal with the justification of most research applications.

Zanakis and Gupta (1985) also claimed that there was a lack of reliable computer codes or software to handle large-scale problems and nonlinear problems. Based on the survey research of computer software presented in Table 3-12 this is no longer the case. The availability of the GP computer software is helping GP literature to grow in two ways. First, the availability of software allows the developers of new methodologies to compare prior methodology with their own newly developed methodology. In reviewing the integer GP (see Tables 3-2 and 3-3), nonlinear GP (see Tables 3-4 and 3-5), interactive GP (see Table 3-6), sequential GP (see Table 3-7) and fuzzy GP (see Table 3-10) methodology citations, half or more of the citations have occurred in just the last ten years (since 1984). In most of these studies some form of software application was utilized to provide a comparative analysis to help confirm the significance of the new GP methodology that was being proposed. A second contribution of GP computer software can be seen in the new applications that are increasingly appearing in the literature. An examination of the dates of GP application citations in the tables in Chapter 4 reveals that over 50 percent of all GP applications occurred since 1984 and virtually all of these studies were run on computer software developed during this time.

Neither GP theory or computer software is lacking in the 1990's. More can always be used, but neither should be used as a reason to keep researchers from doing GP research.

Growing Interest in the Subject

There are a great many optimistic trends that can be observed about GP research. The growth in interest in *multiple criteria decision making* (MCDM) is one positive trend since GP is viewed as a subject within the field (see Figure 1-1). As MCDM grows to increased dominance in the fields of *management science and operations research* (MS/OR), it helps to bring with it increased interest and research opportunities for GP research.

In a survey by the *Operations Research Society of America* and the *Institute of Management Sciences* respondents were asked their opinions about what they thought were the most useful OR/MS methodologies (Dyer 1993). Both professional researchers and recently graduated students were polled in the survey. In the resulting ranking, MCDM was ranked 5th by the professionals and 7th by the recent graduates. The only other category that GP might have fallen into is *mathematical programming*, in which case the professionals ranked this category 6th and the recent graduates ranked it 5th. In either case, this was the first time MCDM had made it into the top ten rankings and a growing sign of GP's relative importance as a decision making methodology.

Another trend that is positioning MCDM for growth in recent years is the increased number of conferences devoted to this subject. The *International Conference on Multiple Criteria Decision Making* which is sponsored by the *International Society on Multiple Criteria Decision Making* held its 12th annual meeting in Hagen, Germany in 1995. The 1st *International Multi-Objective and Goal Programming Theories and Applications* conference took place in Portsmouth, England in 1994. The 1st *International Conference on Multiple Objective Decision Support Systems for Agricultural and Environmental Management* held its conference in Honolulu, Hawaii in 1995.

Perhaps one of the most promising signs of the potential for publications in MCDM occurred when The *Journal of Multi-Criteria Decision Making*, which is published by John Wiley and Sons of New York, started in 1992. In addition to an entire journal being devoted to the subject of MCDM, there have been a variety of international journals that have devoted a special issue to MCDM methods, including GP. In the last five years some of these journals have included *Computers and Operations Research, INFOR, Agricultural Systems* and *Journal of Advanced Transportation*.

Even in the way GP is taught in schools has been the subject of interest and research. Prior research by Hannan (1976) and Lee, Shim and Lee (1984) helped to present basic ideas or strategies for teaching GP to undergraduate students. Recent research on teaching GP by Kennedy (1991), Kim and Kim (1992), and Lee and Kim (1992a, 1992b) is enhancing the understanding of newer *interative GP* methodologies and the use of computer graphic systems. While the potential impact of these newer teaching methods has not yet had a chance to show up in the literature, they will undoubtedly help to increase the interest in GP in a new generations students. According to dissertation research on GP the less recent generation of GP researchers also have quite some interest in GP research (see Lee and Shim 1987a).

SHIFTING THE LIFE CYCLE OF GP RESEARCH TO GROWTH

The *life cycle of GP research* introduced in Chapter 1 and whose depiction is presented in Figure 5-1 appears to portent a rather ominous future trend. The apparent *decline stage* in the life cycle of GP research we find ourselves in can be shifted back into a *maturity* or even *growth stage* by increasing the number of GP publications. But increasing GP publications just for publication sake is not going to lead to a long life. Indeed, it can be argued from a production standpoint, that cheapening a product just to produce more units with less material, will lead to a decline in sales in the long-term because of poor quality.

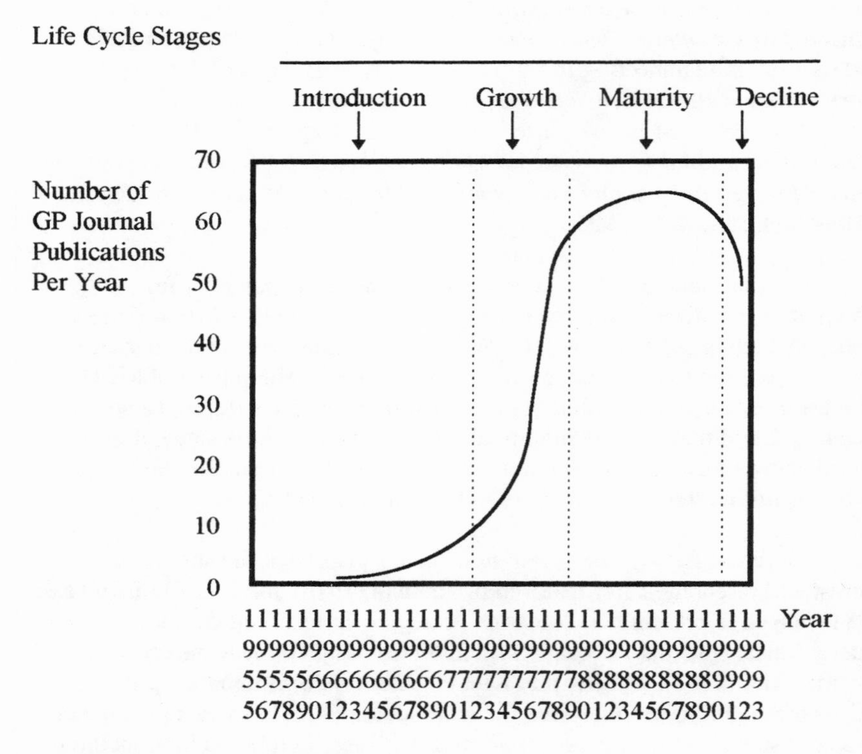

Figure 5-1. Life Cycle of GP Research

To shift the life cycle of GP research back to a growth stage without sacrificing quality can be achieved by valued-added innovation in both methodology and application. Researchers who use linear programming (LP) to model problems still find ways of having their work published when they offer an innovative idea in modeling methodology or application. The same logic can apply to even the most mature GP models. By finding innovative ideas for GP models will lead to value-added, high quality contributions that make a significant contribution to this field of study. The problem is where to find ideas for innovative ideas.

Based on observations of the total body of GP journal research presented in *Appendix B*, a number of possible innovation suggestions can be summarized. These suggested innovative ideas may be known to the experienced GP researcher but are not always articulated or practiced by them. These ideas are particularly meant to challenge and stimulate idea creation for less experienced researchers. Indeed, some of these suggested innovations in GP research may help graduate students come up with an idea for a thesis or dissertation.

1. *New technologies can lead to new applications, particularly in the areas of agriculture, engineering and management*: For example the development of flexible manufacturing systems (FMS) has helped to generate several applications (see Table 4-8, Flexible Manufacturing). New technologies are constantly being invented and introduced into agriculture, engineering and management environments. Each introduction represents a possible area of application. One additional point of opportunity that can be mentioned is the fact that automated systems that reduce variability in the application environment (like a robot replacing a human on an assembly line), improves the use of deterministic methodologies like GP. As variability in production processing is removed, more deterministic modeling methodologies can be applied because of the manufacturing consistency of the automation.

2. *New approaches and philosophy can lead to new GP model opportunities.* For example, the introduction of the production philosophy of *just-in-time* (JIT) management has helped to generate a few new studies using GP modeling (see Table 4-8, Kanban/Just-In-Time) and *total quality management* (TQM) has also generated some research (see Table 4-8, Quality). Considering that there are hundreds of articles on JIT and TQM, the potential for combinations with GP modeling are endless. It should also be remembered that there many other fairly recent and new approaches being introduced that can be used in a similar way in just about every functional area in business.

3. *New government legislation can lead to new GP model opportunities.* For example, the enactment of the *North American Free Trade Agreement* (NAFTA) in 1993 represents one of the greatest change agents to impact business in both North America and the world with which it trades. International contractual economic and environmental impact studies can be under taken using GP. Also, countless studies using management methodologies within the multiple goal framework inherent in NAFTA can be

undertaken on topics including logistics, lot-sizing, maintenance, materials management, project planning, purchasing, quality, sampling, scheduling production and transportation. With the available GP models in Table 4-8, Production and Operations Management as a guide, the opportunities for new NAFTA related GP models are substantial.

4. *Innovations in finance can lead to new GP modeling opportunities.* For example, the development of *junk bonds* for allocating debt was one of the most significant innovations of the 1980's. This finance innovation did not appear in any GP model, even though finance applications related to bonds and mutual funds exist and finance represents the third highest number of GP publications (see Table 4-10). New finance related innovations are being created all the time that could and should be optimized with multiple objectives using GP.

5. *Changes in accounting practices can lead to new GP modeling opportunities.* For example, anyone of the many tax changes each year can be viewed as a possible candidate for a GP model (see Table 4-1, Taxes).

6. *Innovations in marketing practices can lead to new GP modeling opportunities.* For example, consideration of new advertising systems, like the placing of advertisements on the floors of food stores can involve a large number of marketing planning variables. Media mix, layout design and location models that already exist can be reapplied to bring into consideration multiple marketing factors that are not yet considered with this media innovation (see Table 4-9, Media Planning, Retailing, Sales Management).

7. *Seeking better social issue correctness can lead to new GP modeling.* This is an almost undiscovered area (see Table 4-6, Social Issues). For example, military officer accession models (see Table 4-6, Military) can be adjusted to take into consideration the desire for better representation of women and minority groups.

8. *Changes toward international globalism can lead to new GP modeling.* This area more than the others will be one of the most fruitful areas of GP modeling. Previously not identified in other GP bibliographies (Lin 1980, Zanakis and Gupta (1985), Romero (1986, 1991, pp. 96-97), international applications of everything that has been applied nationally represents the greatest possible opportunity for those who are actively involved in research in foreign countries. This includes more than the simple explanations of how something like logistics is done in a foreign country, but how that country and other countries can combine their systems to achieve multiple objectives.

9. *The MS/OR and GP combined strategy (as discussed in Chapter 2) can create new opportunities for GP modeling.* Each time a new MS/OR methodology is developed that can be used in combination with GP, a new potential GP model and application are created. When a new GP methodology is created, the increase in possible publications where that methodology can be combined with all the other MS/OR methodologies is dramatically increased.

10. *Looking on the brighter side of criticism can create new opportunities for GP modeling.* One additional area for research that can be proposed will help undo some of the negative comments and cynicism that exists in the study of GP. The development of examples of GP modeling that counter all the defects in GP modeling that have been reported in the literature over the years is one area of possible future research. Examples that show that GP models don't suffer, or under what situations they may suffer from dominance, inferiority, nonefficiency, naive relative weighting, incommesurability, naive prioritization or redundancy will help others to defend their work from the critics (who are us).

The suggestions for research presented here are a small portion of what is left to be done with GP. It is up to every researchers to find other areas of application. For those who can not find any other innovated areas for GP research, the above suggestions are hereby offered as a challenge to implement.

SUMMARY

This chapter presented a discussion on how GP research is positioned to enter a growth stage in its life cycle of research. Suggestions on how researchers can help contribute to the growth of the research were also presented.

GP is currently positioned well to be a major contributor to the fields of MCDM and MS/OR. There is a more than a sufficient body of knowledge in GP methodology and application to support any use of the model that fits its basic assumptions. There is also sufficient computer software to aid in the application of GP in small, medium and large scale problems solving in industry and academia. There also appears to be a sufficiently growing interest in the subject by students and researchers. Sufficient interest to be enthusiastic about a possible resurgence in GP modeling research and related MCDM methodologies.

It is interesting to note that more than 50 percent of the methodology papers and application papers have been published in the last ten years of the forty year run since the introductory paper by Charnes, Cooper and Ferguson (1955). Interest in GP is not coming to an end, it is slowing down to shift gears for an even greater run. The life cycle of GP research is waiting to be reborn all over again.

REFERENCE

Dyer, J. S., "Suggestions for an OR/MS Master's Degree Curriculum," *OR/MS Today*, Vol. 20, No. 1 (1993), pp. 18-31.

APPENDIX A. TEXTBOOKS, READINGS BOOKS AND MONOGRAPHS ON GOAL PROGRAMMING

The alphabetical by author listing of books in this appendix are those cited in GP research publications and/or having made a significant contribution to GP literature. While some of these books are focused on subjects other than GP, they contain relevant or substantial GP material.

Abadie, J., ed., *Nonlinear Programming*, North-Holland, Amsterdam, 1967.

Chankong, Vira and Haimes, Yacov Y., *Multiobjective Decision Making*, North-Holland Publishing, New York-Amsterdam, 1983.

Charnes, A. and Cooper, W. W., *Management Models and Industrial Applications of Linear Programming*, John Wiley and Sons, New York, NY, 1961.

Cochrane, J. L. and Zeleny, M. eds., *Multiple Criteria Decision Making*, Springer-Verlag, Berlin, 1973.

Cohon, J. L., *Multiobjective Programming and Financial Planning*, Academic Press, New York, NY, 1978.

Depontin, M., Nijkamp, P. and Spronk, J., *Macro-Economic Planning with Conflicting Goals*, Springer-Verlag, Berlin, 1984.

Fandel, G. and Gal, T. eds., *Multiple Objective Decision Making Theory and Applications*, Springer-Verlag, Berlin, 1979.

Fandel, G. and Spronk, J. eds., *Multiple Criteria Decision Methods and Applications*, Springer-Verlag, Berlin, 1985.

Gal, Thomas and Wolf, Hartmut, *Solving Stochastic Linear Programs via Goal Programming*, Springer, Berlin, 1985.

Graves, R. L. and Wolfe, P., eds., *Recent Advances in Mathematical Programming*, McGraw-Hill, New York, NY, 1963.

Haimes, Y. Y. and Chankong, G. V. eds., *Decision Making with Multiple Objectives*, Springer-Verlag, Berlin, 1985.

Hansen, P. ed., *Essays and Surveys on Multiple Criteria Decision Making*, Springer-Verlag, Berlin, 1983.

Hwang, C. L. and Masud, A. S. M. eds., *Multiple Objective Decision Making-Methods and Applications*, Springer, Berlin-Heidelberg, 1979.

Ignizio, James P., *Goal Programming and Extensions*, Heath (Lexington Books), Lexington, MA, 1976.

Ignizio, James P., *Introduction to Linear Goal Programming*, Sage Publications, Beverly Hills, CA, 1985.

Ignizio, James P., *Linear Programming in Single- and Multiple-Objective Systems*, Prentice-Hall, Englewood Cliffs, NJ, 1982.

Ijiri, Y., *Management Goals and Accounting for Control*, Rand-McNally, Chicago, IL, 1965.

Jaaskelainen, V. ed., *Linear Programming and Budgeting*, Petrocelli Books, Princeton, NJ, 1976.

Keeney, R. L. and Raiffa, H., *Decisions with Multiple Objectives: Preference and Value Tradeoffs*, John Wiley & Sons, New York, 1976.

Kwak, N. K. and Schniederjans, M. J., *Introduction to Mathematical Programming*, Kreiger Publishing, Malabar, FL, 1987.

Lawrence, Kenneth D., Guerard, J. B. Jr. and Reeves, G. D. eds., *Advances in Mathematical Programming and Financial Planning*, Vol. 1, JAL Press, Greenwich, CN, 1987.

Lawrence, Kenneth D., Guerard, J. B. Jr. and Reeves, G. D. eds., *Advances in Mathematical Programming and Financial Planning*, Vol. 2, JAL Press, Greenwich, CN, 1990.

Lawrence, Kenneth D., Guerard, J. B. Jr. and Reeves, G. D. eds., *Advances in Mathematical Programming and Financial Planning*, Vol. 3, JAL Press, Greenwich, CN, 1993.

Lee, Sang M. and Van Horn, James, *Academic Administration: Planning, Budgeting and Decision Making by Mutliple Objectives*, University of Nebraska Press, Lincoln, NE, 1983.

Lee, Sang M., *Goal Programming for Decision Analysis*, Auerbach Publishers, Philadelphia, PA, 1972.

Lee, Sang M., *Goal Programming Methods for Multiple Objective Integer Programs*, American Institute of Industrial Engineers, Atlanta, GA, 1979.

Lee, Sang M., *Management by Multiple Objectives*, Petrocelli Books, Princeton, NJ, 1982.

Lieberman, Elliot R., *Multi-Objective Programming in the USSR*, Academic Press, Boston, MA, 1991.

Morse, J. N. ed., *Organizations Multiple Agents with Multiple Criteria*, Springer-Verlag, Berlin, 1981.

Rekanitis, G., Ravindran, A. and Ragsdell, K., *Engineering Optimization: Methods and Applications*, Wiley and Sons, New York, NY, 1983.

Ringuest, Jeffrey L., *Multiobjective Optimization: Behavioral and Computational Considerations*, Kluwer Academic Publishers, Boston, MA, 1992.

Rios-Insua, David, *Sensitivity Analysis in Multi-Objective Decision Making*, Springer-Verlag, Berlin, 1990.

Romero, Carlos, *Handbook of Critical Issues in Goal Programming*, Pergamon Press, Oxford, England, 1991.

Sakawa, Masatoshi, *Fuzzy Sets and Interactive Multiobjective Optimization*, Plenun Press, New York, NY, 1993.

Sawaragi, Yoshikazu, Nakayama, Hirotaka, and Tanino, Tetsuzo, *Theory of Multiobjective Optimization*, Academic Press, Orlando, FL, 1985.

Schniederjans, Marc J., *Linear Goal Programming*, Petrocelli Books, Princeton, NJ, 1984.

Spronk, J., *Interactive Multiple Goal Programming: Applications to Financial Planning*, Martinus Nijhoff, Boston, 1981.

Starr, Martin and Zeleny, M. eds., *Multiple Criteria Decision Making-TIMS Studies in the Management Sciences*, North Holland, Amsterdam, 1977.

Steuer, Ralph E., *Multiple Criteria Optimization: Theory, Computation, and Application*, John Wiley and Sons, New York, NY, 1986.

Van Delft, A. and Nijkamp, P., *Multi-Criteria Analysis and Regional Decision Making, Martinus Nijhoff*, Leiden, The Hague, 1977.

Zeleny, M. ed., *MCDM: Past Decade and Future Trends*, JAI Press, Greewich, CN, 1984.

Zeleny, M. ed., *Multiple Criteria Decision Making. Kyoto*, Springer-Verlag, Berlin, 1975.

Zeleny, M. and Cochrane, J. L., *Multiple Criteria Decision Making*, University of South Carolina Press, Columbia, SC, 1973.

Zoints, S. ed., *Multiple Criteria Problem Solving*, Springer, Berlin-Heidelberg, 1978.

Zoints, S., *Multiple Criteria Mathematical Programming: An Updated Overview and Several Approaches*, Springer, Berlin-New York, 1988.

APPENDIX B. JOURNAL RESEARCH PUBLICATIONS ON GOAL PROGRAMMING

The alphabetical by author listing of journal research publications in this appendix represents the largest collection to date of citations from journals written in English on the subject of GP. While some of these publications are focused on subjects other than GP, they contain relevant or substantial GP material.

Acharya, B., Jain, V. K., and Batra, J. L., "Multi-Objective Optimization of the ECM Process," *Precision Engineering*, Vol. 8, No. 2 (1986), pp. 88-96.

Acharya, B., Nayak, N. N. and Mohanty, D. N., "Dynamic Programming Models with Goal Objectives in Transportation Problems," *Journal of Orissa Mathematical Society*, Vol. 6, No. 2 (1987), pp. 75-88.

Aggarwal, S. C. and Clark, D. J., "A Model for National Planning Policies," *Omega*, Vol. 6, No. 2 (March 1978), pp. 161-171.

Agrawal, U., Swarup, K. and Garg, K. C., "Goal programming Problem with Linear Fractional Objective Function," *Cahiers du Centre-d' Etudes-de Recherche Operationnelle*, Vol. 26, Nos. 1-2 (1984), pp. 33-41.

Akgul, M., "A Note on Lexicographic Linear Programming," *INFOR*, Vol. 22 (1984), p. 343.

Albright, S. C., "Allocation of Research Grants to University Research Proposals," *Socio-Economic Planning Sciences*, Vol. 9 (1975), pp. 189-195.

Alexander, Gordon J., and Resnick, Bruce G., "Using Linear and Goal Programming to Immunize Bond Portfolios," *Journal of Banking and Finance*, Vol. 9, No. 1 (1985), pp. 35-54.

Al-Faraj, T. N., Alidi, A. S. and Al-Ibrahim, A. A., "A Planning Model for Determining the Optimal Location and Size of Traffic Centers: The Case of Damman Metropolitan, Saudi Arabia," *European Journal of Operational Research*, Vol. 66, No. 3 (1993), pp. 272-278.

Alonso, M. A. B. and Devaux, P., "Locations and Size of Day Nurseries--A Mulitple Goal Approach," *European Journal of Operational Research*, Vol. 6, No. 2 (February 1981), pp. 195-198.

Alvord, Charles H. III., "The Pros and Cons of Goal Programming: a Reply," *Computers and Operations Reserach*, Vol. 10, No. 1 (1983), pp. 61-62.

Amador, F. and Romero C., "Redundancy in Lexicographic Goal Programming," *European Journal of Operational Research*, Vol. 41, No. 3 (1989), pp. 347-354.

Anderson, A. M., and Earle, M. D., "Diet Planning in the Third World by Linear and Goal Programming," *Journal of the Operational Research Society*, Vol. 34, No.1 (1983), pp. 9-16.

Armstrong, R. D., Charnes, A. and Haksever, C., "Implementation of Successive Linear Programming Algorithms For Non-Convex Goal Programming," *Computers and Operations Reserach*, Vol. 15, No. 1 (1988), pp. 37-49.

Armstrong, R. D. , Charnes, A. and Haksever, C., "Successive Linear Programming for Ratio Goal Problems," *European Journal of Operational Research*, Vol. 32, No. 3 (1987), pp. 426-434.

Armstrong, R. D. and Cook, W. D., "Goal Programming Models for Assigning Search and Rescue Aircraft to Bases," *Journal of the Operational Research Society*, Vol. 30, No. 6 (June 1979), pp. 555-561.

Armstrong, R. D. and Cook, W. D., "The Contract Formulation Problem in Preventive Pavement Maintenance: A Fixed-Charge Goal Programming Model," *Computers, Environment and Urban Systems*, Vol. 6, No. 2 (1981), pp. 147-155.

Aronson, Jay E. and Thomson, G. L., "The Solution of Multiperiod Personnel Planning Problems by the Forward Simplex Method," *Large Scale Systems: Theory and Applications*, Vol. 9, No. 2 (1985), pp. 129-139.

Arp, Paul A. and Lavigne, R., "Planning with Goal Programming: A Case Study for Multiple-Use of Forested Land," *The Forestry Chronicle*, Vol. 58, No. 5 (1982), pp. 225-232.

Arthur, J. L. and Lawrence, Kenneth D., "A Multiple Goal Capital Flow Model for a Chemical and Pharmaceutical Company," *Engineering Economist*, Vol. 30, No. 2 (1985), pp. 121-134.

Arthur, J. L., and Lawrence, Kenneth D., "A Multiple Goal Blending Problem," *Computers and Operations Reserach*, Vol. 7, No. 3 (1980), pp. 215-224.

Arthur, J. L., and Lawrence, Kenneth D., "Multiple Goal Production and Logistics Planning in a Chemical and Pharmaceutical Company," *Computers and Operations Reserach*, Vol. 9, No. 2 (1982), pp. 127-137.

Arthur, J. L. and Ravindran, A., "An Efficient Goal Programming Algorithm Using Constraint Partitioning and Variable Elimination," *Management Science*, Vol. 24, No. 8 (1978), pp. 867-868.

Arthur, J. L. and Ravindran, A., "A Branch-and-Bound Algorithm with Contraint Partitioning for Integer Goal Programming Problems," *European Journal of Operational Research*, Vol. 4, No. 6 (1980), pp. 421-425.

Arthur, J. L. and Ravindran, A., "A Multiple Objective Nurse Scheduling Model," *AIIE Transactions*, Vol. 13, No. 1 (1981), pp. 55-60.

Arthur, J. L. and Ravindran, A., "Comments on the Special Issue on Generalized Goal Programming," *Computers and Operations Reserach*, Vol. 13, No. 4 (1986), pp. 527-528.

Arthur, J. L. and Ravindran, A., "PAGP: a Partitioning Algorthim for (Linear) Goal Programming Problems," *ACM Tranasactional Mathematical Software*, Vol. 6, No. 3 (1980), pp. 378-386.

Ashton, D. J. and Atkins, D. R., "Multicriteria Programming for Financial Planning," *Journal of the Operational Research Society*, Vol. 30 (1979), pp. 259-270.

Ashton, D. J., "Goal Programming and Intelligent Financial Simulation Models, Part I - Some Problems in Goal Programming," *Accounting and Business Research*, Vol. 16 (Winter 1985), pp. 3-10.

Ashton, D. J., "Goal Programming and Intelligent Financial Simulation Models: Part II - Parametric Searches in Goal Programming," *Accounting and Business Research*, Vol. 16 (Spring 1986), pp. 83-89.

Awerbuch, S., Ecker, J. G. and Wallace, W. A., "A Note: Hidden Nonlinearities in the Application of Goal Programming," *Management Science*, Vol. 22, No. 8 (1976), pp. 918-920.

Badran, Yahya, "Departmental Full Costing via Goal Programming Models," *European Journal of Operational Research*, Vol. 17, No. 3 (1984), pp. 331-337.

Bafail, A. O. and Moreb, A. A., "Optimal Allocation of Students to Different Departments in an Engineering College," *Computers and Industrial Engineering*, Vol. 25, Nos. 1-4 (September 1993), pp. 295-298.

Bailey, A. D., Jr., Boe, W. J. and Schnack, T., "The Audit Staff Assignment Problem: A Comment," *The Accounting Review*, Vol. XLIX, No. 3 (1974), pp. 572-574.

Bajgier, S. M. and Hill, A. V., "An Experimental Comparison of Statistical and Linear Programming Approaches to the Discriminant Problem," *Decision Sciences*, Vol. 13, No. 4 (1982), pp. 604-618.

Baker, J. R. and Fitzpatrick, K. E., "Determination of an Optimal Forecast Model for Ambulance Demand Using Goal Programming," *Journal of the Operational Research Society*, Vol. 37, No. 11 (1986), pp. 1047-1059.

Baker, J. R., Clayton, E. R. and Taylor, B. W. III, "A Non-Linear Multi-criteria Programming Approach for Determining Country Emergency Medical Service Ambulance Allocations," *Journal of the Operational Research Society*, Vol. 40, No. 5 (1989), pp. 423-432.

Baker, J. R., Clayton, E. R. and Moore, L. J., "Redesign of Primary Response Areas For County Ambulance Services," *European Journal of Operational Research*, Vol. 41, No. 1 (1989), pp. 23-32.

Balachandran, K. R. and Steuer, R. E., "An Interactive Model for the CPM Firm Audit Staff Planning Problem with Multiple Objectives," *The Accounting Review*, Vol. LVII, No. 1 (January 1982), pp. 125-140.

Bandyopadhyay, R., "Operational Research in Development Banking in India," *European Journal of Operational Research*, Vol. 2, No. 1 (January 1978), pp. 8-25.

Barbaro, R. W. and Mutmansky, J. M., "Goal Programming Model for Determining the Optimal Production Schedule Considering Penalties or Bonuses Dependent on Quality," *Transactions of the American Institute of Mining, Metallurgical, and Petroleum Engineers*, Vol. 280, Part A (1986), pp. 1887-1895.

Bard, Jonathan F., "A Multiobjective Methodology for Selecting Subsystem Automation Options," *Management Science*, Vol. 32, No. 12 (1986), pp. 1628-1641.

Bard, Jonathan F., "Using Multicriteria Methods in the Early Stages of New Product Development," *Journal of the Operational Research Society*, Vol. 41, No. 8 (1990) pp. 755-766.

Barda, O. H., Dupuis, J. and Lencioni, P., "Multicriteria Location of Thermal Power Plants," *European Journal of Operational Research*, Vol. 45, Nos. 2-3 (1990) pp. 332-346.

Barnett, D., Blake, B. and McCarl, B. A., "Goal Programming via Multidimensional Scaling Applied to Senegalese Subsistence Farms," *American Journal of Agricultural Economics*, Vol. 64, No. 4 (1982), pp. 720-727.

Bartlett, E. T. and Clawson, W. J., "Profit, Meat Production or Efficient Use of Energy in Ranching," *Journal of Animal Science*, Vol. 46, No. 3 (March 1978), pp. 812-818.

Bascaran, Eduardo, Mistree, Farrokh, and Bannerot, Richard B., "Compromise: An Effective Approach for Solving Multiobjective Thermal Design problem," *Engineering Optimization*, Vol. 12 (1987), pp. 175-189.

Basu, M., "A Note on Non-Preemptive Goal Programming," *Journal of Information and Optimization Sciences*, Vol. 5 (1984), pp. 65-67.

Basu, M., Pla, B. B. and Ghosh, D., "The Priority Preferenced Goal Programming Method for Solving Multiobjective Dynamic Programming Models," *Advances in Modelling and Simulation*, Vol. 22, No. 2 (1991), pp. 49-64.

Batson, R. G., "Financial Planning Using Goal Programming," *Long Range Planning*, Vol. 22, No. 5 (1989), pp. 112-120.

Baybars, Iker, "On Currently Practiced Formulations of the Assembly Line Balance Problem," *Journal of Operations Management*, Vol. 5, No. 4 (1985), pp. 449-453.

Bazaraa, M. S. and Bouzaher, A., "A Linear Goal Programming Model for Developing Economies with an Illustration from the Agriculture Sector in Egypt," *Management Science*, Vol. 27, No. 4 (April 1981), pp. 396-413.

Beck, P. O. and Klein, G., "Determining Restrictive Goals in Linear Goal Programs," *Naval Research Logistics Review*, Vol. 36, No. 5 (1989), pp. 675-682.

Beilby, Mary H. and Mott, Thomas H. Jr., "Academic Library Acquisitions Allocation Based on Multiple Collection Development Goals," *Computers and Operations Reserach*, Vol. 10, No. 4 (1983), pp. 335-343.

Ben-Tal, Aharon and Teboulle, Marc, "Expected Utility, Penalty Functions, and Duality in Stochastic Nonlinear Programming," *Management Science*, Vol. 32, No. 11 (1986), pp. 1445- 1466.

Benayoun, R., De Montgolfier, J., Tergny, J. and Laritchev, D., "Linear Programming with Multiple Objective Functions: Step Method (STEM)," *Mathematical Programming*, Vol. 1, No. 3 (1971), pp. 366-375.

Benito, M. A. and Devaux, P., "Location and Size of Day Nurseries - A Multiple Goal Approach," *European Journal of Operational Research*, Vol. 6, No. 2 (February 1981), pp. 195-198.

Benjamin, C. O., "A Linear Goal-Programming Model for Public-Sector Project Selection," *Journal of the Operational Research Society*, Vol. 36, No.1 (1985), pp. 13-23.

Benjamin, C. O., Ehie, Ike C. and Omurtag, Y., "Planning Facilities at the University of Missouri-Rolla," *Interfaces*, Vol. 22, No. 4 (July/August 1992), pp. 95-105.

Berlo, J. M. van, "A Decision Support Tool for the Vegetable Processing Industry: An Integrative Approach to Market , Industry and Agriculture," *Agriculture Systems*, Vol. 43, No. 1 (1993), pp. 91-109.

Bhattacharya, U., Rao, J. R. and Tiwari, R. N., "Bi-criteria Multi-Facility Location Problem in Fuzzy Environment," *Fuzzy Sets and Systems*, Vol. 56, No. 2 (1993), pp. 145-153.

Bhattacharya, U., Rao, J. R. and Tiwari, R. N., "Fuzzy Multi-criteria Facility Location Problem," *Fuzzy Sets and Systems*, Vol. 51, No. 3 (1992), pp. 277-287.

Bishop, A B., Narayanan, R., Grenney, W. J. and Pugner, P. E., "Goal Programming Model for Water Quality Planning," *Journal of the Environmental Engineering* Division, ASCE, EE2 (1977), pp. 293-305.

Bit, A. K. and Alam, S. S., "Optimal Planning for Allocation of Coal Energy by Goal Programming," *Industrial Engineering Journal*, Vol. 22, No. 6 (1993), pp. 8-12.

Bit, A. K. Biswal, M. P. and Alam, S. S., "An Additive Fuzzy Programming Model for Multiobjective Transportation Problem," *Fuzzy Sets and Systems*, Vol. 57, No. 3 (1993), pp. 313-319.

Blocher, E., "Sampling for Integrated Audit Objective - A Comment," *The Accounting Review*, Vol. 53 (July 1978), pp. 766-774.

Bonczek, R. H., Holsapple, C. W. and Whinston, A. B., "Computer-Based Support of Organizational Decision Making," *Decision Sciences*, Vol. 10, No. 2 (1979), pp. 268-291.

Booth, G. G. and Bessler, W., "Goal Programming Models for Managing Interest-Rate Risk," *Omega*, Vol. 17, No. 1 (1989), pp. 81-92.

Booth, G. G. and Dash, G. H. Jr., "Alternate Programming Structures for Bank Portfolios," *Journal of Banking and Finance*, Vol. 3, No. 1 (1979), pp. 67-82.

Booth, G. G. and Dash, G. H. Jr., "Bank Portfolio Management Using Non-Linear Goal Programming," *The Finance Review*, Vol. 12 (1977), pp. 59-69.

Bottoms, K. E. and Bartlett, E. T., "Resource Allocation through Goal Programming," *Journal of Range Management*, Vol. 28, No. 6 (1975), pp. 442-447.

Bouzaher, A. and Mendoza, G. A., "Goal Programming: Potential and Limitations for Agricultural Economics," *Canadian Journal of Agricultlural Economics*, Vol. 35 (March 1987), pp. 89-107.

Brauer, D. C. and Naadimuth, G., "A Goal Programming Model for Aggregate Inventory and Distribution Planning," *Mathematical and Computer Modelling*, Vol. 16, No. 3 (1992), pp. 81-90.

Bres, E. S., Burns, D. Charnes, A. and Cooper, W. W., "A Goal Programming Model for Planning Officer Accessions," *Management Science*, Vol. 26, No. 8 (August 1980), pp. 773-783.

Brown, K and Norgaard, R., "Modelling the Telecommunication Pricing Decision," *Decision Sciences*, Vol. 23, No. 3 (1992), pp. 673-686.

Brown, L. G., McClendon, R. W. and Akbay, K. S., "Goal Programming for Estimating Initial Influx of Insect Pests," *Agricultural Systems*, Vol. 34, No. 4 (1990), pp. 337-348.

Buchanan, J. T. and Daellenbach, H. G., "A Comparative Evaluation of Interactive Solution Methods for Multiple Objective Decision Models," *European Journal of Operational Research*, Vol. 29, No. 3 (1987), pp. 353-359.

Buchanan, J. T., "Multiple Objective Mathematical Programming: A Review," *New Zealand Operational Research*, Vol. 14, No. 1 (1986), pp. 1-27.

Buckley, J. J., "Compatibility of Multiple Goal Programming and the Maximize Expected Utility Criterion," *Theory and Decision*, Vol. 16, No. 4 (1984), pp. 209-216.

Budavei, V., "The Program-Goal Method of National Economic Planning," *International Studies of Management and Organization*, Vol. 11, Nos. 3-4 (Fall/Winter 1982), pp. 142-156.

Buffa, Frank P. and Jackson, Wade M., "A Goal Programming Model for Purchase Planning," *Journal of Purchasing and Materials Management*, Vol. 19, No. 3 (1983), pp. 27-34.

Buffa, Frank P. and Shearon, W. T., "Goal Programming for Allocating Resources in a Municipal Recreation Department," *Journal of Leisure Research*, Vol. 12, No. 2 (1980), pp. 128-137.

Buffa, Frank P., "A Goal Programming Model for Simulataneous Determination of Safety Stock Levels," *Production and Inventory Management*, Vol. 17, No. 1 (1976), pp. 94-104.

Buffa, Frank P., "A Model for Allocating Limited Resources when Making Safety-Stock Decisions," *Decision Sciences*, Vol. 8, No. 2 (April 1977), pp. 415-426.

Buffa, Frank P., "A Zero-One Budgeting Process with Goal Programming Feedback," *Computers, Environment and Urban Systems*, Vol. 8 (1983), pp. 93-108.

Buhl, H. U., "Axiomatic Considerations in Multi-objective Location Theory," *European Journal of Operational Research*, Vol. 37, No. 3 (1988), pp. 363-367.

Buongiorno, J. and Svanqvist, N., "A Seperable Goal Programming Model of the Indonesian Forestry Sector," *Forest Ecologic Management*, Vol. 4, No. 1 (January 1982), pp. 67-78.

Butler, T. W., Karwan, K. R., Sweigart, J. R. and Reeves, G. R., "An Integrateive Model-Based Approach to Hospital Layout," *IIE Transactions*, Vol. 24, No. 2 (May 1992), pp. 144-152.

Callahan, J. R., "An Introduction to Financial Planning Through Goal Programming," *Cost and Management*, Vol. 3, No. 1 (January-February 1973), pp. 7-12.

Campbell, H. and Ignizio, J. P., "Using Linear Programming for Predicting Student Performance," *Journal of Education and Psychological Measurement*, Vol. 32 (1972), pp. 397-401.

Can E. K. and Houck, M. H., "Real-time Reservior Operations by Goal Programming," *Journal of Water Resources Planning and Management*, Vol. 110 (1984), pp. 297-309.

Caplin, D. A. and Kornbluth, J. S. H., "Multiobjective Investment Planning Under Certainty," *Omega*, Vol. 3, No. 4 (1975), pp. 423-441.

Carlsson, C., "Tackling an MCDM-Problem with the Help of Some Results from Fuzzy Set Theory," *European Journal of Operational Research*, Vol. 10, No. 3 (1982), pp. 270-281.

Cavalier, T. M., Ignizio, J. P. and Soyster, A. L., "Discriminant Analysis via Mathematical Programming: Certain Problems and Their Causes," *Computer and Operations Research*, Vol. 16, No. 4 (1989), pp. 353-362.

Chae, Young M., Newbrander, W. C. and Thomason, J. A., "Application of Goal Programming to Improve Resource Allocation for Health Services in Papua New Guinea," *International Journal of Health Planning and Management*, Vol. 4, No. 2 (April-June 1989), pp. 81-95.

Chae, Young M., Suver, James D., and Chou, David, "Goal Programming as a Capital Investment Tool for Teaching Hospitals," *Health Care Management Review*, Vol. 10, No. 1 (1985), pp. 27-35.

Chakraborty, T. K., "A Single Sampling Attribute Plan of Given Strength Based on Fuzzy Goal Programming," *Operations Reseach*, Vol. 25, No. 4 (December 1988a), pp. 259-271.

Chakraborty, T. K., "A Preemptive Single Sampling Attribute Plan of Given Strength," *Opsearch*, Vol. 23, No. 3 (1986), pp. 164-174.

Chakraborty, T. K., "Fuzzy Goal Programming Approach for Designing Single Sampling Attribute Plans When Sample Size is Fixed," *IAPQR Transactions*, Vol. 16, No. 2 (1991a), pp. 1-8.

Chakraborty, T. K., "Goal Programming Approach for Designing a Single Sampling Attribute Plans When Sample Size is Fixed," *IAPQR Transactions*, Vol. 13, No. 2 (1988b), pp. 19-30.

Chakraborty, T. K., "Stochastic Parameter Single Sampling Plans of Given Strength," *Calcutta Statistical Association Bulletin*, Vol. 41, Nos. 161-164 (1991b), pp. 117-126.

Chambers, D. and Charnes, A., "Inter-Temporal Analysis and Optimization of Bank Portfolios," *Management Science*, Vol. 7 (1961), pp. 393-410.

Chanda, E. C. K., "Application of Goal Programming to Production Planning in the Crushed Stone Industry," *International Journal of Surface Mining and Reclamation*, Vol. 4, No. 3 (1990), pp. 125-129.

Chandler, J. S., "A Multiple Criteria Approach for Evaluting Information Systems," *MIS Quaterly*, Vol. 6 (1982), pp. 61-74.

Changchit, C. and Terrell, M. P., "A Multiobjective Reservoir Operation Model with Stochastic Inflows," *Computers and Industrial Engineering*, Vol. 24, No. 2 (1993), pp. 303-313.

Chang, D. T., Chung, D. S. and Hwang, C. L., "Nonlinear Goal Programming for Optimizing Rice Conditioning and Storage Systems: Part I," *Journal of the Korean Society of Agriculatural Machinery*, Vol. 8, No. 2 (December 1983), pp. 69-85.

Chang, D. T., Chung, D. S. and Hwang, C. L., "Nonlinear Goal Programming for Optimizing Rice Conditioning and Storage Systems: Part II," *Journal of the Korean Society of Agriculatural Machinery*, Vol. 9, No. 1 (June 1984), pp. 46-52.

Chang, Peter and Perl, Jossef, "A Finite Element Based Approach To Multi-Objective Structural Optimization Using Goal Programming," *Engineering Optimization*, Vol. 13 (1988), pp. 65-82.

Chang, S. J. and Buongiorno, J., "A Programming Model for Multiple Use Forestry," *Journal of Environmental Management*, Vol. 13 (1981), pp. 41-54.

Charnes, A. and Collomb, B., "Optimal Economic Stabilization Policy: Linear Goal-Interval Programming Models," *Socio-Economic Planning Sciences*, Vol. 6, No. 4 (1972), pp. 431-435.

Charnes, A. and Cooper, W. W., "Goal Programming and Constrained Regression--A Comment," *Omega*, Vol. 3, No. 4 (1975), pp. 403-409.

Charnes, A. and Cooper, W. W., "Goal Programming and Multiple Objective Optimization (Part 1)," *European Journal of Operational Research*, Vol. 1, No. 1 (January 1977), pp. 39-54.

Charnes, A. and Cooper, W. W., "Programming with Linear Fractional Functionals," *Naval Research Logistics Quarterly*, Vol. 9 (1962), pp. 181-186.

Charnes, A. and Stedry, A. C., "Search-Theoretic Models of Organization Control by Budgeted Multiple Goals," *Management Science*, Vol. 12, No. 5 (1966), pp. 457-481.

Charnes, A. and Storbeck, J. "A Goal Programming Model For the Citing Multilevel EMS Systems," *Socio-Economic Planning Sciences*, Vol. 14, No. 4 (1980), pp. 155-161.

Charnes, A., Colantoni, C. and Cooper, W. W., "A Futurological Justification for Historical Cost and Multi-Dimensional Accounting," *Accounting, Organizations and Society*, Vol. 1, No. 4 (1976), pp. 315-337.

Charnes, A., Colantoni, C., Cooper, W. W. and Kortanek, K. O., "Economic Social and Enterprise Accounting and Mathematical Models," *The Accounting Review*, Vol. 47 No. 1 (1972), pp. 85-108.

Charnes, A., Cooper, W. W. and Ferguson, R. O., "Optimal Estimation of Executive Compensation by Linear Programming," *Management Science*, Vol. 1, No. 2 (1955), pp. 138-151.

Charnes, A., Cooper, W. W. and Ijiri, Y., "Breakeven Budgeting and Programming to Goals," *Journal of Accounting Research*, Vol. 1, No. 1 (Spring 1963), pp. 16-43.

Charnes, A., Cooper, W. W. and Sueyoshi, T., "A Goal Programming/Constrained Regression Review of The Bell System breakup," *Management Science*, Vol. 34, No. 1 (1988), pp. 1-26.

Charnes, A., Cooper, W. W. and Sueyoshi, T., "Least Squares/Ridge Regression and Goal Programming/ Constrained Regression Alternatives," *European Journal of Operational Research*, Vol. 27, No. 2 (1986), pp. 146-157.

Charnes, A., Cooper, W. W., DeVoe, J. K., Learner, D. B. and Reinecke, W., "A Goal Programming Model for Media Planning," *Management Science*, Vol. 14, No. 8 (1968), pp. B423-B430.

Charnes, A., Cooper, W. W., Harrald, J., Karwan, K. R. and Wallace, W. A., "A Goal Interval Programming Model for Resource Allocation in a Marine Environmental Economics and Management," *Journal of Environmental Economics and Management*, Vol. 3, No. 4 (1976), pp. 347-362.

Charnes, A., Cooper, W. W., Karwan, K. and Wallace, W. A., "A Chance-Constrained Goal Programming Model to Evaluate Response Resources for Marine Pollution Disasters," *Journal of Environmental Economics and Management*, Vol. 6, No. 3 (1979), pp. 244-274.

Charnes, A., Cooper, W. W., Klingman, D. and Niehaus, R. J., "Explicit Solutions in Convex Goal Programming," *Management Science*, Vol. 22, No. 4 (1975), pp. 438-448.

Charnes, A., Cooper, W. W., Learner, D. B. and Snow, E. F., "Note on an Application of Goal Programming Model for Media Planning," *Management Science*, Vol. 14, No. 8 (April 1968), pp. B431-B436.

Charnes, A., Cooper, W. W., Niehaus, R. J. and Stedry, A., "Static and Dynamic Assignment Models with Multiple Objectives, and Some Remarks on Organization Design," *Management Science*, Vol. 15, No. 8 (1969), pp. B365-B375.

Charnes, A., Cooper, W. W. and Niehaus, R. J., "Dynamic Multi-Attribute Models for Mixed Manpower Systems," *Naval Research Logistics Quaterly*, Vol. 22, No. 2 (1975), pp. 205-220.

Charnes, A., Duffuaa, S. and Al-Saffar, A., "A Dynamic Goal Programming Model for Planning Food Self-Sufficiency in the Middle East," *Applied Mathematical Modeling*, Vol. 13, No. 2 (February 1989), pp. 86-93.

Charnes, A., Duffuaa, S. and Intriligator, M., "Extremal and Game Theorectic Characterizations of the Probabilistic Approach to Income Redistribution," *Journal of Optimization Theory and Applications*, Vol. 44, No. 3 (1984), pp. 435-451.

Charnes, A., Haynes, K. E., Hazleton, J. E. and Ryan, M. J., "A Hierarchical Goal-Programming Approach to Environmental Land Use Management," *Geographical Analysis*, Vol. 7, No. 2 (1975), pp. 121-130.

Chateau, Jean-Pierre D., "The Capital Budgeting Problem Under Conflicting Financial Policies," *Journal of Business Finance and Accounting*, Vol. 2, No. 1 (Spring 1975), pp. 83-103.

Chaudhry, S. S., Forst, F. G. and Zydiak, J. L., "A Multicriteria Approach to Allocating Order Quantity Amoung Vendors," *Production and Inventory Management Journal*, Vol. 32, No. 3 (1991), pp. 82-86.

Chen, J. G. and Yeung T. W., "Hybrid Expert System Approach to Nurse Scheduling," *Computers and Nursing*, Vol. 11, No. 4 (1993), pp. 183-190.

Chen, Joyce T., "A Simplified Integer Programming Approach to Resource Allocation and Profit Budgeting," *Accounting and Business Research*, Vol. 13, No. 52 (1983), pp. 273-278.

Chen, Ling-Hwei, Farn, Kwo-Jean and Tsay, Ching-Shu, "A Distributed Data Allocation Algorithm Based on the Zero-One Goal Programming," *Journal of the Chinese Institute of Chemical Engineers*, Vol. 14, No. 5 (1991), pp. 551+.

Chen, Yung-Jung and Askin, R.G., "A Multi Objective Evaluation of Flexible Manufacturing System Loading Heuristics," *International Journal of Production Research*, Vol. 28, No. 3 (1990) pp. 895-912.

Chen, Y. X., "A Chance-Constrained Goal Programming Model and Its Application to the Decision for Risky Investments," *Systems Engineering*, Vol. 5, No. 4 (1987), pp. 24-31.

Chen, Y. X., "Multiple Objective Levelling Programming and Its Applications to the Regional Planning with Multi-Counties," *Systems Engineering*, Vol. 4, No. 4 (1986), pp. 12-17.

Chen, Y. X., "Multiple Resource-Constrained Multi-Period Scheduling: A Zero-One Goal Programming Model and Its Application," *Systems Engineering*, Vol. 6, No. 2 (1988), pp. 17-24.

Chetty Mallikarjuna, K. and Subramanian, D. K., "Rural Energy Consumption Patterns with Multiple Objectives," *International Journal of Energy Research*, Vol. 12, No. 3 (1988), pp. 561-567.

Chicoine, D., Scott, J. T. Jr. and Jones, T. W., "The Application of Goal Programming in Rural Land Use Policy," *Journal of the Community Development Society*, Vol. 11, No. 1 (Spring 1980), pp. 77-94.

Chisman, J. A. and Rippy, D., "Optimal Operation of a Multipurpose Reservior Using Goal Programming," *Review of Industrial Management and Textile Science*, (Fall 1977), pp. 69-82.

Choi, T. S. and Levary, R. R., "Multi-National Capital Budgeting Using Chance-Constrained Goal Programming," *International Journal of Systems Science*, Vol. 20, No. 3 (March 1989), pp. 395-414

Choo, Eng U. and Wesley, William C., "Optimal Criterion Weights in Repetitive Multicriteria Decision Making," *Journal of the Operational Research Society*, Vol. 36, No. 11 (1985), pp. 983-992.

Choobineh, Fred and Burgman, R., "Transmission Line Route Selection: An Application of K-Shortest Paths and Goal Programming," *IEEE Transactions on Power Apparatus and Systems*, VPAS-103 (1984), pp. 3253-3259.

Christainson, J. B., "Balancing Policy Objectives in Long-term Care," *Health Service Research*, Vol. 13 (1983),pp. 157-170.

Clayton, E. R. and Moore, L. J., "An Interactive Model for Advertising Media Selection," *Southern Journal of Business*, Vol. 7, No. 4 (1972a), pp. 37-45.

Clayton, E. R. and Moore, L. J., "Goal vs. Linear Programming," *Journal of Systems Management*, Vol. 23, No. 11 (November 1972b), pp. 26-31.

Clayton, E. R., Weber, W. E. and Taylor, B. W. III, "A Goal Programming Approach to the Optimization of Multiresponse Simulation Models," *AIIE Transactions*, Vol. 14, No. 4 (1982), pp. 282-287.

Cocklin, C., Lonergan, S. C. and Smit, B., "Assessing Options in Resource Use for Renewable Energy Through Multiobjective Goal Programming," *Environment and Planning Analysis*, Vol. 18, No. 10 (October 1986), pp. 1323-1338.

Cocks, K. D. and Baird, I. A., "Using Mathematical Programming to Address the Multiple Reserve Selection Problem: An Example From the Eyre Peninsula, South Australia," *Biological Conservation*, Vol. 49, No. 2 (1989), pp. 113-130.

Cohn, E. and Morgan, J. M., "Improving Resource Allocation Within School Districts: A Goal Programming Approach," *Journal of Education Finance*, Vol. 4, No. 1 (Summer 1978), pp. 89-104.

Cohon, J. L. and Marks, D. H., "A Review and Evaluation of Multiobjective Programming Techniques," *Water Reources Research*, Vol. 11, No. 2 (1975), pp. 208-220.

Collons, R. W., Gass, S. I. and Rosendahl, E. E., " The ASCAR Model for Evaluating Military Manpower Policy," *Interfaces*, Vol. 13, No. 3 (1983), pp. 44-53.

Contini, B., "A Stochastic Approach to Goal Programming," *Operations Reseach*, Vol. 16, No. 3 (1968), pp. 576-586.

Cook, Wade D. and Kress, M., "Deriving Weights from Pairwise Comparison Ratio Matricies: An Axiomatic Approach," *European Journal of Operational Research*, Vol. 37, No. 3 (1988), pp. 355-362

Cook, Wade D., "Goal Programming and Financial Planning Models for Highway Rehabilitation," *Journal of the Operational Research Society*, Vol. 35, No. 3 (1984), pp. 217-223.

Cook, Wade D., "Zero-Sum Games with Multiple Goals," *Naval Research Logistics Quarterly*, Vol. 23, No. 4 (1976), pp. 615-622.

Cornett, D. and Williams, W. A., "Goal Programming for Multiple Land Use Planning at Mineral King, California," *Journal of Soil and Water Conservation*, Vol. 46 (September-October 1991), pp. 373-376.

Courtney, J. F. Jr., Klastorian, T. D. and Ruefli, T. W., "A Goal Programming Approach to Urban-Surburban Location Preferences," *Management Science*, Vol. 18, No. 6 (1972), pp. B258-B268.

Croucher, J. S., "A Goal Programming Model for Timetable Scheduling," *Opsearch*, Vol. 21 (1984), pp. 145-152.

Crowder, Lee J. and Sposito, V. A., "Comments on 'Algorithm for Solving the Linear Goal Programming Problem by Solving Its Dual," *Journal of the Operational Research Society*, Vol. 38, No. 4 (1987), pp. 335-340.

Crowder, Lee J. and Sposito, V. A., "Sequential linear Goal Programming : Implementation via MPSX/370E," *Computers and Operations Reserach*, Vol. 18, No.3 (1991) pp. 291-296.

Cubbage, F. W., Field, R. C., Eza, D. A. and Farkas, A., "Determining Optimal Forest Land Ownership Patterns: A Goal Programming Approach," *Resource Management Optimization*, Vol. 6, No. 1 (1987), pp. 1-14.

Current, J., Min, H. and D. Schilling, "Multi Objective Analysis of Facility Location Decisions," *European Journal of Operational Research*, Vol. 49, No. 3 (1990) pp. 295-308.

Czuchra, W., "Interative Goal Programming Approach to Sharing Resources Amoung Dependent Operations," *Foundations of Control Engineering*, Vol. 10, No. 3 (1985), pp. 113-122.

Czuchra, W., "Multicriteria Discrete Resource Allocation Among Dependent Operations by Interative Goal Programming Method," *Systems Science*, Vol. 14, No. 1 (1988), pp. 39-46.

Dane, C. W., Meador, N. C. and White, J. B., "Goal Programming in Land-Use Planning," *Journal of Forestry*, Vol. 75, No. 6 (June 1977), pp. 325-329.

Daniels, R.L., "A Multi-Objective Approach to Resource Allocation in Single Machine Scheduling," *European Journal of Operational Research*, Vol. 48, No. 2 (1990) pp. 226-241.

Darby-Dowman, K. and Mitra, G., "An Extension of Set Partitioning with Application to Scheduling Problems," *European Journal of Operational Research*, Vol. 21, No. 2 (1985), pp. 200-205.

Dauer, J. P., "Analysis of the Objective Space in Multiple Objective Linear Programming," *Journal of Mathematical Analysis and Applications*, Vol. 126, No. 2 (1987), pp. 579-598.

Dauer, J. P. and Krueger, R. J., "A Multiobjective Optimization Model for Water Resources Planing," *Applied Mathematical Modelling*, Vol. 4, No. 3 (1980), pp. 171-175.

Dauer, J. P. and Krueger, R. J., "Mathematical Programming Techniques in Majorization," *Journal of Optimization Theory and Applications*, Vol. 25, No. 3 (1978), pp. 361-373.

Dauer, J. P. and Krueger, R. J., "An Iterative Approach to Goal Programming," *Operational Research Quaterly*, Vol. 28, No. 3 (1977), pp. 671-681.

Dauer, J. P. and Lim, Y. H., "Solving Multiple Objective Linear Programs in Objective Space," *European Journal of Operational Research*, Vol. 46, No. 3 (1990) pp. 350-357.

Davis, K. R., Stam, A. and Grzybowski, R. A., "Resource Constrained Project Scheduling with Multiple Objectives: A Decision Support System," *Computers and Operations Research*, Vol. 19, No. 7 (1992), pp. 657-669.

Davis, W. and Talavage, J., "Three-Level Models for Hierarchical Coordination," *Omega*, Vol. 5, No. 6 (1977), pp. 709-720.

Davis, W. and Whitford, D. T., "A Note on the Optimality of the Modified Generalized Goal Decomposition Model," *Management Science*, Vol. 31, No. 5 (1985), pp. 640-643.

De, P. K., Acharya, D. and Sahu, K. C., "A Chance-Constrained Goal Programming Model for Capital Budgeting," *Journal of the Operational Research Society*, Vol. 33 (July 1982), pp. 635-638.

De, P. K., Acharya, D. and Sahu, K. C., "Goal Programming Model for Capital Budgeting Under Risk," *Journal of the Institution of Engineers* (India), Part IDP, Vol. 67, Part 1 (1986), pp. 18-20.

Dean, B. V. and Schniederjans, Marc J., "A Multiple Objective Selection Methodology for Strategic Industry Selection Analysis," *IEEE Transactions*, Vol. 38, No. 1 (1991), pp. 53-62.

Dean, B. V., Schniederjans, Marc J. and Yu, Y. M., "A Goal Programming Approach to Production Planning for Flexible Manufacturing Systems," *Journal of Engineering and Technology Management*, Vol. 1, No. 6 (1990), pp. 207-220.

Deckro, Richard F. and Hebert, John E., "A Mulitple Objective Programming Framework for Tradeooffs in Project Scheduling,"*Engineering Costs and Production Economics*, Vol. 18, No. 3 (1990), pp.255-264.

Deckro, Richard F. and Hebert, John E., "Goal Programming Approaches to Solving Linear Decision Rule Based Aggregate Production Planning Models," *AIIE Transactions*, Vol. 16, No. 4 (1984), pp. 308-315.

Deckro, Richard F. and Hebert, John E., "Polynomial Goal Programming: A Procedure for Modeling Preference Tradeoff," *Journal of Operations Management*, Vol. 7, No. 2 (December 1988), pp. 149-164.

Deckro, R. F. and Rangachari, S., "A Goal Approach to Assembly Line Balancing," *Computers and Operations Research*, Vol. 17, No. 5 (1990), pp. 509-521.

Deckro, Richard F., Hebert, John E. and Winkofsky, E. P., "Mulitple Criteria Job-Shop Scheduling," *Computers and Operations Reserach*, Vol. 9, No. 4 (1984), pp. 279-285.

Deckro, Richard F., Spahr, Ronald W. and Hebert, John E., "Preference Trade-Offs in Capital Budgeting Decisions," *AIIE Transaction*, Vol. 17, No.4 (1985), pp. 332-337.

De Kluyver, Cornelis A., "A Generalized Upper Bound Algorithm for Large-Scale Generalized Goal Programming Problems," *International Journal of Systems Science*, Vol. 8, No. 8 (1977), pp. 883-904.

De Kluyver, Cornelis A., "An Exploration of Various Goal Programming Formulations--With Aplication to Advertising Media Selection," *Journal of the Operational Research Society*, Vol. 30, No. 2 (February 1979a), pp. 167-171.

De Kluyver, Cornelis A. and Moskowitz, Herbert, "Assessing Scenario Probabilities Via Interactive Goal Programming," *Management Science* Vol. 30, No. 3 (1984), pp. 273-278.

De Kluyver, Cornelis A., "Hard and Soft Constraints in Media Scheduling," *Journal of Advertising Research*, Vol. 18 (1978), pp. 27-31.

De Kluyver, Cornelis A., "On the Importance of Goal-Norming in Non-Preemptive Goal Programming," *Opsearch*, Vol. 16, Nos. 3-4 (1979b), pp. 88-97.

DePorter, Elden L. and Ellis, Kimberly P., "Optimization of Project Networks with Goal Programming and Fuzzy Linear Programming," *Computers and Industrial Engineering*, Vol. 19, No. 1-4 (1990) pp. 500-504.

Dessent, G. and Hume, B., "Value for Money and Prison Perimeters-Goal Programming for Goals," *Journal of the Operational Research Society*, Vol. 14, No. 7 (July 1990), pp. 583-590.

De Wit, C. T., Van Keulen, H., Seligman, N. G. and Spharim, I., "Application of Interactive Multiple Goal Programming Techniques for Analysis and Planning of Regional Agricultural Development," *Agricultural Systems*, Vol. 26, No. 3 (1988), pp. 211-230.

Dieperink, Han and Nijkamp, Peter, "Multiple Creteria Location Model for Innovative Firms in a Communication Network," *Economic Geography*, Vol. 63, No. 1 (1987), pp 66-73.

Diminnie, Carol B. and Kwak, N. K., "A Hierarchical Goal-Programming Approach to Reverse Resource Allocation in Institutions of Higher Learning," *Journal of the Operational Research Society*, Vol. 37, No. 1 (1986), pp. 59-66.

Dinklebach, Werner, "On Nonlinear Fractional Programming," *Management Science*, Vol. 13, No. 7 (1967), pp. 422-498.

Djang, P. A., "Selecting Personal Computers," *Journal of Research on Computing Education*, Vol. 25, No. 3 (1993), pp. 327-338.

Dobbins, G. L. and Mapp, H. P. Jr., "A Comparison of Objective Function Structures Used in a Recursive Goal Programming Simulations Model of Farm Growth," *Southern Journal of Agricultural Economics*, Vol. 14, No. 2 (1982), pp. 9-16.

Donckels, R., "Regional Multiobjective Planning Under Uncertainty: A Stochastic Goal Programming Formulation," *Journal of Regional Science*, Vol. 17, No. 2 (1977), pp. 207-216.

Drandell, M., "A Resource Association Model for Insurance Management Utilizing Goal Programming," *Journal of Risk and Insurance*, Vol. 44, No. 2 (1977), pp. 311-315.

Drezner, Z. and Wesolowsky, G.O., "Design of Multiple Criteria Sampling Plans and Charts," *International Journal of Production Research*, Vol. 29, No. 1 (1991) pp. 155-163.

Drynan, Ross G. and Sandiford, Frences, "Incorporating Economic Objectives in Goal Programs For Fishery Management," *Marine Resource Economics*, Vol. 2, No. 2 (1985), pp. 175-195.

Drynan, Ross G., "Goal Programming and Multiple Criteria Decision Making in Farm Planning: An Expository Analysis-A Comment," *Journal of Agricultural Economics*, Vol. 36, No. 3 (1985), pp. 421-423.

Dusansky, R. and Kalman, D. J., "Regional Multi-Objective Planning Under Uncertainty," *Regional Science and Urban Economics*, Vol. 11, No. 1 (1981), pp. 121-134.

Dyer, A. A., Hof, J. G., Kelly, J. W., Crim, S. A. and Alward, G. S., "Implications of Goal Programming in Forest Resource Allocation," *Forest Science*, Vol. 25, No. 4 (1979), pp. 535-543.

Dyer, A. A., Hof, J. G., Kelly, J. W., Crim, S. A. and Alward, G. S., "Implications of Goal Programming in Forest Resource Allocation: A Reply," *Forest Science*, Vol. 29, No. 4 (1983), pp. 837-840.

Dyer, James S., "A Time-Sharing Computer Program for the Solution of the Multiple Criteria Problem," *Management Science*, Vol. 19, No. 12 (1973), pp. 1379-1383.

Dyer, James S., "Interactive Goal Programming," *Management Science*, Vol. 19, No. 1 (1972), pp. 62-70.

Dyer, R. F., Forman, E. H. and Mustafa, M. A., "Decision Support for Media Selection Using the Analytic Hierarchy Process," *Journal of Advertising*, Vol. 21, No. 1 (March 1992), pp. 59-72.

Eatman, J. L. and Sealey, C. W. Jr., "A Multiobjective Linear Programming Model for Commercial Bank Balance Sheet Management," *Jounal of Bank Research*, Vol. 9, No. 4 (1979), pp. 227-236.

Ebrahimpour, M. and Ansari, A., "Measuring the Effectiveness of Quality Control Circles: A Goal Programming Approach," *International Journal of Operations and Production Management*, Vol. 8, No. 2 (1988), pp. 59-68.

Ehir, Ike C. and Benjamin, C. O., "An Integrated Multiobjective Planning Model: A Case Study of the Zambian Copper Mining Industry," *European Journal of Operational Research*, Vol. 68, No. 2 (July 1993), pp. 160-172.

Eilon, Samuel, "Goals and Constraints in Decision Making," *Operational Research Quaterly*, Vol. 23, No. 1 (1972), pp. 3-16.

Eilon, Samuel, "Goals and Constraints," *Journal of Management Studies*, Vol. 8, No. 3 (1971), pp. 292-303.

Eilon, Samuel, "Multi-Criteria Warehouse Location," *International Journal of Physical Distribution and Materials Management*, Vol. 12, No. 1 (1982), pp. 42-45.

El-Dash, A. A. and Mohamed, M.B., "Sequential Duality Method for Solving Polynomial Goal Programming Problems," *Eqyptian Computer Journal*, Vol. 20, No. 1 (1992), pp. 12-38.

El-Dash, A. A., "Two Probabilistic Models for Solving an Oxygen-Bottle Inventory Problem," *Journal of the Operational Research Society*, Vol. 40, No. 11 (November 1989), pp. 961-969.

El-Sayed, M. E. M., Ridgely, B. J. and Sandgren, E., "Nonlinear Structural Optimization Using Goal Programming," *Computers and Structures*, Vol. 32, No. 1 (1989), pp. 69-73.

El-Sheshai, K. M., Harwood, G. B. and Hermanson, R. H., "Cost Volume Profit Analysis with Integer Goal Programming," *Management Accounting*, Vol. LIX, No. 4 (1977), pp. 43-47.

El-Shishiny, H., "A Goal Programming Model for Planning and Development of Newly Reclaimed Lands," *Agricultural Systems*, Vol. 26, No. 4 (1988), pp. 245-261.

Elamin, I. M., Duffuaa, S. O. and Yassein, H. A., "Transmission Line Route Selection by Goal Programming," *International Journal of Electrical Power and Energy Systems*, Vol. 12, No. 2 (1990), pp. 138-143.

Eom, Hyun B. and Lee, Sang M., "A Large-Scale Goal Programming Model Based Decision Support for Formulating Global Financing Strategy," *Information and Management*, Vol. 12, No. 1 (January 1987), pp. 33-44.

Eom, Hyun B., Lee, Sang M., Snyder, C. A. and Ford, F. N., "A Multiple Criteria Decision Support System for Global Financing Strategy," *Financial Planning*, Vol. 4, No. 3 (Winter 1987-88), pp. 94-113.

Eto, H., "A Mathematical Programming Model to Consider Alternative Crops," *Omega*, Vol. 19, Nos. 2-3 (1991), pp. 169-179.

Evans, Davis S. and Heckman, James J., "Rejoinder: Natural Monopoly and the Bell System: Response to Charnes, Cooper, and Sueyoshi," *Management Science*, Vol. 34, No. 1 (1988), pp. 27-38.

Evans, J. P. and Steuer, R. E., "A Revised Simplex Method for Linear Multiple Objective Programs," *Mathematical Programming*, Vol. 5 (1973), pp. 54-72.

Everitt, R. R., Sonntag, N. C., Puterman, M. L. and Whalen, P., "A Mathematical Programming Model for the Management of a Renewable Resource System: The Kemano II Development Project," *Journal of the Fisheries Research Board of Canada*, Vol. 35 (1978), pp. 235-246.

Fahmy, D. and El-Shishiny, H., "A Goal Programming Model for Desert Farm Planning," *Advanced Desert Arid Land Technology Development*, Vol. 5 (1991), pp. 69-86.

Ferreira, P. A. V. and Geromel, J. C., "An Interactive Projection Method for Multicriteria Optimization problems," *IEEE Transactions on Systems, Man, and Cybernetics*, Vol. 20, No. 3 (1990) pp. 596-605.

Feuer, M. J., "Organizational Decline, Extended Work Life and Implications for Faculty Planning," *Socio-Economic Planning Sciences*, Vol. 19, No. 3 (1985), pp. 213-221.

Fichefet, J., "GPSTEM: A Interactive Multi-Objective Optimization Method," *Progress in Operations Research*, Vol. 1 (1976), pp. 317-332.

Field, D. B., "Goal Programming for Forest Management," *Forest Science*, Vol. 19, No. 2 (1973), pp. 125-135.

Field, R. C., Dress, P. E. and Fortson, J. C., "Complementary Linear and Goal Programming Procedures for Timber Harvest Scheduling," *Forest Science*, Vol. 26, No. 1 (1980), pp. 121-133.

Filios Vassilios P., "Social Accountability and its Auditing," *Socio-Economic Planning Sciences*, Vol. 18 No. 2 (1984), pp. 117-125.

Fisk, J. C., "A Goal Programming Model for Output Planning," *Decision Sciences*, Vol. 10, No. 4 (1979), pp. 593-603.

Fisk, J. C., "Production Scheduling for High-Volume Assemblies - A Case Study," *Journal of the Operational Research Society*, Vol. 31, No. 9 (1980), pp. 781-789.

Fishburn, P. C., "Lecicographic Orders, Utilities and Decision Rules: A Survey," *Management Science*, Vol. 20 (1974), pp. 1442-1471.

Fisher, G. W., Wei, Y. and Dontamsetti, S., "Process-Controlled machining of Gary Cast Iron," *Journal of Mechanical Working Technology*, Vol. 20 (September 1989), pp. 47-57.

Flavelli, R. B., "A New Goal Programming Formulation," *Omega*, Vol. 4, No. 6 (1976), pp. 731-732.

Flinn, J. C., Jayasuriya, S. and Knight, C. G., "Incorporating Multiple Objectives in Planning Models of Low-Resource Farmers," *The Australian Journal of Agricultural Economics*, Vol. 24, No. 1 (1980), pp. 35-45.

Forsyth, J. D., "Utilization of Goal Programming in Production and Capital Expenditure Planning," *CORS Journal*, Vol. 7, No. 2 (1969), pp. 136-140.

Fortenberry, Jessie C., Mitra, Amitava and Willis, R. D. M., "A Multi-Criteria Approach to Optimal Emergency Vehicle Location Analysis," *Computers and Industrial Engineering*, Vol. 16, No. 2 (1989) pp. 339-347.

Fortson, J. C. and Dince, R. R., "An Application of Goal Programming to Management of a Country Bank," *Journal of Bank Reserach*, Vol. 7, No. 4 (1977), pp. 311-319.

Franz, Lori S., Baker, H. M., Leong, G. K. and Rakes, Terry R., "A Mathematical Model for Scheduling and Staffing Multiclinic Health Regions," *European Journal of Operational Research*, Vol. 41 (1989), pp. 227-289.

Franz, Lori S., "Data Driven Modeling: An Application in Scheduling," *Decision Sciences*, Vol. 20, No. 2 (1989), pp. 359-377.

Franz, Lori S., Lee, Sang M. and Van Horn, J. C., "An Adaptive Decision Support System for Academic Resource Planning, *Decision Sciences*, Vol. 12, No. 2 (1981), pp. 276-293.

Franz Lori S., Rakes, Terry R., and Wynne, A. James, "A Chance-Constrained Multiobjective Model for Mental Health Services Planning," *Socio-Economic Planning Sciences*, Vol. 18, No. 2 (1984), pp. 89-95.

Frazier, S. K., Gaither, N. and Olson, D., "A Procedure for Dealing with Multiple Objectives in Cell Formation Decisions," *Journal of Operations Management*, Vol. 9, No. 4 (1990), pp. 465-480.

Freed, Ned and Glover, Fred, "A Linear Programming Approach to the Discriminant Problem," *Decision Sciences*, Vol. 12, No. 1 (January 1981a), pp. 68-74.

Freed, Ned and Glover, Fred, "Evaluating Alternative Linear Programming Models to Solve the Two-Group Discriminant Problem," *Decision Sciences* Vol. 17, No. 2 (1986a), pp. 151-162.

Freed, Ned and Glover, Fred, "Resolving Certain Difficulties and Improving the Classification Power of LP Discriminant Analysis Formulations," *Decision Sciences*, Vol. 17, No.4 (1986b), pp. 589-595.

Freed, Ned and Glover, Fred, "Simple But Powerful Goal Programming Models for Discriminant Problems," *European Journal of Operational Research*, Vol. 7, No. 1 (1981b), pp. 44-60.

Freeland, J. R., "A Note on Goal Decomposition in a Decentralized Organization," *Management Science*, Vol. 23, No. 1 (1976), pp. 100-102.

Freeland, J. R. and Baker, N. R., "Goal Partitioning in a Heirarchical Organization," *Omega*, Vol. 3, No. 6 (1975), pp. 673-688.

Fukukawa, Tadaaki and Hong, Sung-Chan, "The Determination of the Optimal Number of Kanbans in a Just-In-Time Production System," *Computers and Industrial Engineering*, Vol. 24, No. 4 (October 1993), pp. 551-559.

Gabbani, D. and Magazine, M., "An Interactive Heuristic Approach for Multi-Objective Integer Programming Problems," *Journal of the Operational Research Society*, Vol. 37, No. 3 (March 1986), pp. 285-291.

Gallagher, M. A. and Kelly, E. J., "A New Methodology for Military Force Structure Analysis," *Operations Reseach*, Vol. 39, No. 6 (November/December 1991), pp. 877-885.

Gal, Thomas and Wolf, Karin, "Stability in Vector Maximization - A Survey," *European Journal of Operational Research*, Vol. 25, No. 2 (1986), pp. 169-182.

Gangan, S., Khator, S. K. and Bahu, A.J. G., "Multiobjective Decision Making Approach for Determining Alternative Routing in a Flexible Manufacturing System," *Computers and Industrial Engineering*, Vol. 13, Nos. 1-4 (March 1987), pp. 112-117.

Gardner, John.C., Huefner, R.J. and Lotfi, Vahid, "A Multiperiod Audit Staff Planning Model Using Multiple Objectives : Development and Evaluations," *Decision Sciences*, Vol. 21, No. 1 (Winter 1990), pp. 154-170.

Garrod, N. W. and Moores, B., "An Implicit Enumeration Algorithm for Solving Zero-One Goal Programming Problems, *Omega*, Vol. 6, No. 4 (1978), pp. 374-377.

Garrod, N. W., "The Integrative Role of Analytical Decision Models in Public Enterprizes," *Public Enterprize*, Vol. 11, No. 1 (March 1991), pp. 55-63.

Gass, Saul I., "A Process for Determining Priorities and Weights for Large-Scale Linear Goal Programmes," *Journal of the Operational Research Society*, Vol. 37, No. 8 (1986), pp. 779-785.

Gass, Saul I. and Dror, M., "An Interactive Approach to Multiple - objective Linear Programming Involving Key Decision Variables," *Large Scale Systems*, Vol. 64 (1983), pp. 95-103.

Gass, Saul I., Collins, R. W., Meinhardt, C. W., Lemon, D. M. and Gillette, M. D., "The Army Manpower Long-Range Planning System," *Operations Research*, Vol. 36, No. 1 (1988), pp. 5-17.

Gass, Saul I., "Military Manpower Planning Models," *Computers and Operations Research*, Vol. 18, No. 1 (1991), pp. 65-74.

Gass, Saul I., "The Setting of Weights in Linear Goal- Programming Problems," *Computers and Operations Reserach*, Vol. 14, No. 3 (1987), pp. 227-230.

Gembicki, F. and Haimes, Y. Y., "Approach to Performance and Multiobjective Sensitivity Optimization: The Goal Attainment Method," *IEEE Transactions on Automatic Control*, AC-20, No. 1 (1975), pp. 769-771.

Gen, M., Ida, K., Tsujimura, Y. and Chang, E. K., "Large-Scale 0-1 Fuzzy Goal Programming and Its Application to Reliability Optimization Problem," *Computers and Industrial Engineering*, Vol. 24, No. 4 (1993), pp. 539-549.

Gen, M., Ida, K. and Lee, J. U., "A Computational Algorithm for Solving 0-1 Goal Programming with GUB Structures and Its Application for Optimization Problems of a System Reliability," *Electronics and Communications in Japan*, Part 3, Vol. 73, No. 12 (December 1990), pp. 88-96.

Gen, M., Ida, K. and Lee, J. U., "Algorithm for Solving Large-Scale 0-1 Goal Programming and Its Application to Reliability Optimization Problems," *Computers and Industrial Engineering*, Vol. 17, Nos. 1-4 (March 1989), pp. 525-530.

Gen, M., Ida, K. and Xiaojun, Z., "A Method for Solving Goal Programming Problems by Iterative PFL Technique," *Transactions of the Institute of Electronics, Information and Communication Engineers*, Vol. J75-A, No. 10 (October 1992), pp. 1596-1599.

Geoffrion, A. M., Dyer, J. S. and Fienberg, A., "An Interactive Approach for Multi-Criterion Optimization, with an Application to the Operation of an Academic Department," *Management Science*, Vol. 19, No. 4 (1972), pp. 357-368.

Ghandfaroush, P., "Optimal Allocation of Time in a Hospital Pharmacy Using Goal Programming," *European Journal of Operational Research*, Vol. 70, No. 2 (1993), pp. 191-198.

Ghosh, D., Paul, B. B. and Basu, M., "Implementation of Goal Programming in Long-Range Resource Planning in University Management," *Optimization*, Vol. 24, Nos. 3-4 (1992), pp. 373-383.

Ghosh, D., Paul, B. B. and Basu, M., "Determination of Optimal Land Allocation in Agriculatural Planning Through Goal Programming with Penalty Functions," *Opsearch*, Vol. 30, No. 1 (1993), pp. 15-34.

Gibbs, T. E., "Goal Programming," *Journal of Systems Management*, Vol. 24, No. 5 (May 1973), pp. 38-41.

Gibson, Michael, Bernardo, J. J., Chung, C. and Badinelli, R., "A Comparison of Interactive Multiple-Objective Decision Making Procedures," *Computers and Operations Reserach*, Vol. 14, No. 2 (1987), pp. 97-105.

Gilgeous, Vic, "Modeling Realism in Aggregate Planning: A Goal Search Approach," *International Journal of Production Reserach*, Vol. 27, No. 7 (July 1989), pp. 1179-1193.

Gingrich, Gerry and Soli, Sigfrid, D., "Subjective Evaluation and Allocation of Resources in Routine Decision Making," *Organizational Behavior and Human Performance*, Vol. 33 (April 1984), pp. 187-203.

Giokas, D. and Vassiloglou, M., "A Goal Programming Model for Bank Assets and Liabilities Management," *European Journal of Operational Research*, Vol. 50, No. 1 (1991) pp. 48-60.

Gleason, J. M. and Lilly, C. C., "A Goal Programming Model for Insurance Agency Management," *Decision Sciences*, Vol. 8, No. 1 (January 1977), pp. 180-190.

Goicoechea, A., Duckstein, L. and Fogel, M. M., "Multiobjective Programming in Watershed Management: A Study of the Charleston Watershed," *Water Resource Research*, Vol. 12, No. 6 (1976), pp. 1085-1092.

Golany, B., Yadin, M. and Learner, O., "A Goal Programming Inventory Control Model Applied at a Large Chemical Plant," *Production and Inventory Management Journal*, Vol. 32, No. 1 (1991), pp. 16-24.

Golovanov, O. V., Zotov, E. A., Maikov, G. P., Pushnyak, V. A. and Tolstykh, A. V., "Improving Goal-Programmed Management in the Instrumentation Industry," *Automation and Remote Control*, Vol. 48, No. 4, Part 1 (1987), pp. 477-483.

Goodman, D. A., "A Goal Programming Approach to Aggregate Planning and Work Force," *Management Science*, Vol. 20, No. 12 (1974), pp. 1569-1575.

Green, G. I., Kim, C. S. and Lee, Sang M., "A Multi-Criteria Warehouse Location Model," *International Journal of Physical Distribution and Materials Management*, Vol. 11, No. 1 (1981), pp. 5-13.

Green, M. K., McCarthy, P. and Pearl, M., "Multi-Objective Allocation," *Omega*, Vol. 11, No. 2 (1983), pp. 195-200.

Greenwood, Allen G. and Moore, Laurence J., "An Inter-Temporal Multi-Goal Linear Goal Programming for Optimizing University Tuition and fee Structure," *Journal of the Operational Research Society*, Vol. 38, No. 7 (1987), pp. 599-614.

Gross, J. and Talavage, J., "A Multiple-Objective Planning Methodlogy for Information Service Managers," *Information Processing and Management*, Vol. 15, No. 3 (1979), pp. 155-167.

Grove, M. A., "A Surrogate for Linear Programs with Random Requirements," *European Journal of Operational Research*, Vol. 34, No. 3 (1988), pp. 399-402.

Grove, M. A., "Certainty Equivalents for Goal Programs," *Omega*, Vol. 13, No. 4 (1985), pp. 361-362.

Gunther, R. E., Johnson, G. D. and Peterson, R. S., "Currently Practiced Formulations for the Assembly Line Balance Problem," *Journal of Operations Management*, Vol. 3, No. 4 (1983), pp. 209-221.

Gupta, A. K. and Sharma, J. K., "Integer Quadratic Goal Programming," *Journal of the Institution of Engineers* (India), Part PR: Production Engineering Division, Vol. 70, No. 2 (1989), pp. 43-47.

Gupta, Jatinder N.D. and Ahmed, Nazim U., "A Goal Programming Approach to Job Evaluation," *Computers and Industrial Engineering*, Vol. 14, No. 2 (1988) pp. 147-152.

Guven, H. M., Mistree, Farrokh and Bannerot, R. B., "Design Synthesis of Parabolic Through Solar Collectors for Developing Countries," *Engineering Optimization*, Vol. 7, No. 3 (1984), pp. 173-194.

Habeeb, Y. A., "Adapting Multi-criteria Planning to the Nigerian Economy," *Journal of the Operational Research Society*, Vol. 42, No. 10 (October 1991), pp. 885-888.

Haimes, Y. Y. and Li, D., "Hierarchical Multiobjective Analysis for Large-Scale Systems: Review and Current Status," *Automatica*, Vol. 24, No. 1 (January 1988), pp. 53-69.

Hallefjord, A. and Jornsten, K., "A Critical Comment on Integer Goal Programming," *Journal of the Operational Research Society*, Vol. 39, No. 1 (1988), pp. 101-104.

Hannan, Edward L., "An Assessment of Some Criticisms of Goal Programming," *Computers and Operations Reserach*, Vol. 12 No. 6 (1985), pp. 525-541.

Hannan, Edward L., "A Graphical Interpretation in Goal Programming Problems," *Omega*, Vol. 4, No. 6 (1976), pp. 733-735.

Hannan, Edward L., "Allocation of Library Funds for Books and Standing Orders--A Multiple Objective Formulation," *Computers and Operations Research*, Vol. 5, No. 2 (1978a), pp. 109-14.

Hannan, Edward L. and Narasimhan, R., " On Fuzzy Goal Programming - Some Comments/Some Further Comments on Fuzzy Priorities," *Decision Sciences*, Vol. 12, No. 3 (1981), pp. 522-541.

Hannan, Edward L., "Contrasting Fuzzy Goal Programming and "Fuzzy" Multicriteria Programming," *Decision Sciences*, Vol. 13, No. 2 (April 1982a), pp. 337-339.

Hannan, Edward L., "Effects of Substituting a Linear Goal for a Fractional Goal in the Goal Programming Problem," *Management Science*, Vol. 24, No. 1 (1977), pp. 105-107.

Hannan, Edward L., "Linear Programming with Multiple Fuzzy Goals," *Fuzzy Sets and Systems*, Vol. 6, No. 3 (1981a), pp. 235-248.

Hannan, Edward L., "Nondominance in Goal Programming," *INFOR*, Vol. 18, No. 4 (1980), pp. 300-309.

Hannan, Edward L., "Notes on--An Interpretation of Fractional Objectives in Goal Programming as Related to Papers by Awerbuch et al., and Hannan," *Management Science*, Vol. 27, No. 7 (July 1981b), pp. 847-848.

Hannan, Edward L., "On Fuzzy Goal Programming," *Decision Sciences*, Vol. 12, No. 3 (1981c), pp. 522-531.

Hannan, Edward L., "Reformulating Zero-One Games with Multiple Goals," *Naval Research Logistics Quarterly*, Vol. 29, No. 1 (1982b), pp. 113-118.

Hannan, Edward L., "Some Further Comments on Fuzzy Priorities," *Decision Sciences*, Vol. 12, No. 3 (1981d), pp. 539-541.

Hannan, Edward L., "The Application of Goal Programming Techniques to the CPM Problem," *Socio-Economic Planning Sciences*, Vol. 12, No. 5 (1978b), pp. 267-270.

Hansen, B., "A New Tool: Goal Programming [in the Production and Marketing of Christman Trees]," *American Christmas Tree Journal*, Vol. 22, No. 3 (1978), pp. 42-44.

Harrald, J., Leotta, J., Wallace, W. A. and Wendell, R. E., "A Note on the Limitations of Goal Programming as Observed in Resource Allocation for Marine Environmental Protection," *Naval Research Logistics Quaterly*, Vol. 25, No. 4 (1978), pp. 733-739.

Harrington, T. C. and Fischer, W. A., "Portfolio Modeling in Multiple-Criteria Situtations Under Uncertainty: Comment," *Decision Sciences*, Vol. 11, No. 1 (January 1980), pp. 171-177.

Harwood, G. B. and Lawless, R. W., "Optimizing Organizational Goals in Assigning Faculty Teaching Schedules," *Decision Sciences*, Vol. 6, No. 3 (1975), pp. 513-524.

Hattenschwiler, P., "Goal Programming Becomes Most Useful Using L1-Smoothing Functions," *Computational Statistics and Data Analysis*, Vol. 6. No. 4 (1988), pp. 369-384.

Hawkins, C. A. and Adams, R. A., "A Goal Programming Model for Capital Budgeting," *Financial Management*, Vol. 3, No. 1 (Spring 1974), pp. 52-57.

Hemaida, R. S. and Kwak, N. K., "A Linear Goal Programming Model for Trans-Shipment Problems with Flexible Supply and Demand Constraints," *Journal of the Operational Research Society*, Vol. 45, No. 2 (February 1994), pp. 215-224.

Henderson, John C. and Schilling, David A., "Design and Implementation of Decision Support Systems in the Public Sector," *MIS Quaterly*, Vol. 9, No. 2 (1985), pp. 157-169.

Henderson, John C., "Integrated Approach for Manpower Planning in the Service Sector," *Omega*, Vol. 10, No. 1 (1982), pp. 61-73.

Hendrix, G. G. and Stedry, A. C., "The Elementary Redundancy-Optimization Problem: A Case Study in Probabilistic Multiple Goal Programming," *Operations Research*, Vol. 22, No. 3 (1974), pp. 639-653.

Hershauer, J. C. and Gowens, J. W., "Machine Scheduling with Mixed Integer Goal Programming," *Omega*, Vol. 5, No. 5 (1977), pp. 609-610.

Hibiki, N. and Fukukawa, T., "Goal Programming Model Approach for Risk Management in Banking Based on Asset Liability Management (ALM)," *Journal of the Operations Reseach Society of Japan*, Vol. 35, No. 4 (December 1992), pp. 319-344.

Hill, M. and Werczberger, E., "Goal Programming and the Goals Achievement Matrix," *International Regional Science Review*, Vol. 3 (1978), pp. 165-181.

Hindelang, T. J. and Hill, J. L., "A New Model for Aggregate Output Planning," *Omega*, Vol. 6, No. 3 (May 1978), pp. 267-272.

Hindelang, T. J., "QC Optimization Through Goal Programming," *Quality Progress*, Vol. 6, No. 12 (December 1973), pp. 20-22.

Hoffman, James J. and Schniederjans, Marc J., "An International Strategic Management/Goal Programming Model for Structuring Global Expansion Decisions in the Hospitality Industry: The Case Study of Eastern Europe," *International Journal of Hospitality Management*, Vol. 9, No. 3 (1990), pp. 175-190.

Hoffman, James J., Schniederjans, Marc J. and Sirmans, G. Stacy, "A Multi-Criteria Model for Corporate Property Evaluation," *The Journal of Real Estate Research*, Vol. 5, No. 3 (Fall 1990), pp. 285-300.

Hollis, M. S. and Murray, L. W., "Multinational Banking Exchange Risk Strategies: A Support Tool," *Management International Review*, Vol. 25, No. 2 (1985), pp. 5-11.

Hollis, M. S., "Short-Term Foreign Risk Management: Zero Net Exposure Models," *Omega*, Vol. 6 (1978), pp. 249-256.

Hong, H. K., "Finance Mix and Capital Structure," Journal of *Business Finance and Accounting*," Vol. 8 (1981), pp. 485-491.

Hotvedt, J. E., "Application of Linear Goal Programming to Forest Harvest Scheduling," *Southern Journal of Agricultural Economics*, Vol. 15, No. 1 (1983), pp. 103-108.

Hotvedt, J. E., Leushner, W. A. and Buhyoff, "A Heuristic Weight Determination for Goal Programs Used for Harvest Scheduling Models," *Canadian Journal of Forestry*, Vol. 12 (1982), pp. 292-298.

Houck, M. H., "Designing an Expert System for Real-Time Reservior System Operation," *Civil Engineering Systems*, Vol. 2 (1985), pp. 30-37.

Hrubes, Robert J. and Rensi, G., "Implications of Goal Programming in Forest Resource Allocation: Some Comments," *Forest Science*, Vol. 27, No. 3 (1981), pp. 454-459.

Hsu, J. I. S., "Goal Programming Approach to Investment Decision Making," *Marquette Business Review*, Vol 20, No. 4 (1976), pp. 166-170.

Hussein, M. L., "Qualitative Analysis of Basic Notions in an Iterative Approach to Probabilistic Goal Programming," *Fuzzy Sets and Systems*, Vol. 54, No. 1 (1993), pp. 39-46.

Hwang, C. L., Lee, H. B. , Tillman, F. A. and Lie, C. H., "Nonlinear Integer Goal Programming Applied to Optimal System Reliability," *IEEE Transactions on Reliability*, VR-33 (1984), pp. 431-438.

Hwang, C. L., Paidy, S. R., Yoon, K. and Masud, A. S. M., "Mathematical Programming with Multiple Objectives: A Tutorial," *Computers and Operations Reserach*, Vol. 7, No. 1 (1980), pp. 5-31.

Ichida, K. and Fujii, Y., "Multicriterion Optimization Using Interval Analysis," *Computing* (Vienna/New York), Vol. 44, No. 1 (1990), pp. 47-57.

Ignizio, James P., "A Generalized Goal Programming Approach to the Minimal Interface, Multicriteria N + 1 Scheduling Problem," *AIIE Transactions*, Vol. 16, No. 4 (1984a), pp. 316-322.

Ignizio, James P., "An Approach to the Capital Budgeting Problem with Multiple Objectives," *The Engineering Economist*, Vol. 21, No. 4 (1976), pp. 259-272.

Ignizio, James P., "A Note on Computational Methods in Lexicographic Linear Goal Programming," *Journal of the Operational Research Society*, Vol. 34, No. 6 (June 1983a), pp. 539-542.

Ignizio, James P., "A Note on the Multidimensional Dual," *European Journal of Operational Research*, Vol. 17, No. 1 (July 1984b), pp. 116-122.

Ignizio, James P., "A Reply to 'Comments on an Algorithm for Solving the Linear Goal Programming Problem by Solving Its Dual," *Journal of the Operational Research Society*, Vol. 38, No. 12 (1987), pp. 1149-1154.

Ignizio, James P., "A Reply to: "Comments on the Special Issue on Generalized Goal Programming," *Computers and Operations Reserach*, Vol. 13, No. 4 (1986), pp. 528-529.

Ignizio, James P., "A Review of Goal Programming: A Tool for Multi-Objective Analysis," *Journal of the Operational Research Society*, Vol. 29, No.11 (1978a), pp. 1109-1119.

Ignizio, James P., "An Algorithm for Solving the Linear Goal Programming Problem by Solving Its Dual," *Journal of the Operational Research Society*, Vol. 36, No. 6 (1985a), pp. 507-515.

Ignizio, James P., "An Approach to the Modeling and Analysis of Multiobjective Generalized Networks," *European Journal of Operational Research*, Vol. 12, No. 4 (April 1983b), pp. 357-361.

Ignizio, James P. and Daniels, S. C., "Fuzzy Multicriteria Integer Programming via Fuzzy Generalized Networks," *Fuzzy Sets and Systems*, Vol. 10 (1983), pp. 261-270.

Ignizio, James P. and Hannan, E. L., "On the (Re)Discovery of Fuzzy Goal Programming/Contrasting Fuzzy Goal Programming and "Fuzzy" Multicriteria Programming," *Decision Sciences*, Vol. 13, No. 2 (1982), pp. 331-339.

Ignizio, James P. and Perlis, J. H., "Sequential Linear Goal Programming: Implementation via MPSX," *Computers and Operations Reserach*, Vol. 6, No. 3 (1979), pp. 141-145.

Ignizio, James P. and Satterfield, D. E., "Antenna Array Bean Pattern Synthesis via Goal Programming," *Military Electronics Defense*, (September 1977), pp. 402-417.

Ignizio, James P. and Thomas, L. C., "An Enhanced Conversion Scheme for Lexicographic Multiobjective Integer Programming," *European Journal of Operational Research*, Vol. 18 (1984), pp. 57-61.

Ignizio, James P., "An Introduction to Goal Programming with Applications to Urban Systems," *Computers, Environment and Urban Systems*, Vol. 5, No. 1 (1980), pp. 15-33.

Ignizio, James P., "Antenna Array Beam Pattern Synthesis via Goal Programming," *European Journal of Operational Research*, Vo. 6, No. 3 (March 1981a), pp. 286-290.

Ignizio, James P., "Comments on the Special Volume: Mathematical Programming with Multiple Objectives," *Computers and Operations Reserach*, Vol. 8, No. 4 (1981b), pp. 355-356.

Ignizio, James P., "Generalized Goal Programming: An Overview," *Computers and Operations Reserach*, Vol. 10, No. 4 (1983), pp. 277-289.

Ignizio, James P., "Goal Aggregation Via Shadow Prices-Some Counterexamples': A Reply," *Large Scale Systems*, Vol. 12, No. 1 (1987), pp. 87-88.

Ignizio, James P., "GP-GN: An Approach to Certain Large Scale Multiobjective Integer Programming Models," *Large Scale Systems*, Vol. 4 (1983c), pp. 177-188.

Ignizio, James P., "Integer GP via Goal Aggregation," *Large Scale Systems*, Vol. 8 (1985b), pp. 81-86.

Ignizio, James P., "Multiobjective Mathematical Programming via the MULTIPLEX Model and Algorithm," *European Journal of Operational Research*, Vol.22, No.3 (1985c), pp.338-346.

Ignizio, James P., "On the Merits and Demerits of Integer Goal Programming," *Journal of the Operational Research Society*, Vol. 40, No. 8 (August 1989), pp. 781-785.

Ignizio, James P., Palmer, D. F. and Murphy, C. M., "A Multicriteria Approach to the Overal Design of Distributed Computing Systems," *IEEE Transactions on Computers*, C-31 (1982), pp. 410-418.

Ignizio, James P., "The Determination of a Subset of Efficient Solutions via Goal Programming," *Computers and Operations Reserach*, Vol. 8, No. 1 (1981c), pp. 9-16.

Ignizio, James P., "The Development of Cost Estimating Relationships via Goal Programming," *The Engineering Economist*, Vol. 24, No. 1 (Fall 1978b), pp. 37-47.

Ignizio, James P., Wiemann, K. W. and Hughes, J., "Sonar Array Element: A Hybrid Expert Systems Application," *European Journal of Operational Research*, Vol. 32 (1987), pp. 76-85.

Ijiri, Y. and Kaplan, R. S., "A Model for Integrating Sampling Objectives in Auditing," *Journal of Accounting Research*, Vol. 9 (1971), pp. 73-87.

Imany Mohammad M. and Schlesinger, Robert J., "Decision Models for Robot Selection: A Comparison of Ordinary Least Squares and Linear Goal Programming Methods," *Decision Sciences*, Vol. 20, No. 1 (Winter 1989), pp. 40-53.

Inuiguchi, Masahiro and Yosufumi, Kume, "Goal Programming Problems with Interval Coefficients and Target Intervals," *European Journal of Operational Research*, Vol. 52, No. 3 (1991) pp. 345-360.

Irani, S. A., Mittal, R. O. and Lehtihet, E. A., "Tolerance Chart Optimization," *International Journal of Production Research*, Vol. 27, No. 9 (September 1989), pp. 1531-1552.

Iserman, H., "Linear Lexicographic Optimization," *OR Spectrum*, Vol. 4 (1982), pp. 223-228.

Ishibuchi, H. and Tanaka, H., "Multiobjective Programming in Optimization of the Interval Objective Function," *European Journal of Operational Research*, Vol. 48, No. 2 (1990) pp. 219-226.

Ishiyama, A., Hondoh, M., Ishida, M. and Onuki, T., "Optimal Design of MRI Magnets with Magnetic Shielding," *IEEE Transactions on Magnetics*, Vol. 25, No. 2 (1989), pp. 1885-1888.

Jaaskelainen, V. and Lee, S. M., "A Goal Programming Model for Financial Planning," *The Finnish Journal of Business Economics*, Vol. 3, No. 2 (1971), pp. 291-303.

Jaaskelainen, V., "A Goal Programming of Aggregate Production Planning," *Swedish Journal of Economics*, Vol. 71, No. 1 (1969), pp. 14-29.

Jaaskelainen, V., "Strategic Planning with Goal Programming," *Management Information*, Vol. 1, No. 1 (1972), pp. 23-31.

Jackman, H. W., "Financing Public Hospitals in Ontario: A Case Study in Rationing of Capital Budgeting," *Management Science*, Vol. 20, No. 4 (1973), pp. 645-655.

Jacobs, Timothy L. and Wright, Jeff R., "Optimal Inter-Process Steel Production Scheduling," *Computers and Operations Reserach*, Vol. 15, No. 6 (1988), pp. 497-507.

Jain, Hemant K. and Dutta, Amitava, "Distributed Computer System Design: A Multicriteria Decision-Making Methodology," *Decision Sciences*, Vol. 17, No. 4 (1986), pp. 437-453.

Jain, Hemant K., "Comprehensive Model for the Storage Structure Design of Codasyl Databases," *Information Systems*, Vol. 9 (1984), pp. 217-230.

Jain, S. K., Soni, B. and Seethapathi, P. V., "Optimization Technique for Water Resources Management," *Journal of the Institute of Engineers* (India), Part CI: Civil Engineering Division, Vol. 69, Part 1 (July 1988), pp. 16-19.

Jandy, G. and Tanczos, K., "Network Scheduling Limited by Special Constraint as a Function of Time Cost," *Periodica Polytechnica Transportation Engineering*, Vol. 15, No. 2 (1987), pp. 111-123.

Jedrzejowics, P. and Rosicka, L., "Multicriterial Reliability Optimization Problem," *Foundations of Control Engineering*, Vol. 8 (1983), pp. 165-173.

Johnson, J. H., "Applying Goal Programming to Multi-Plant/Product Aggregate Production Loading," *Western Electric Engineering*, (October 1976), pp. 8-15.

Johnson, P. J., Oltenacu, P. A., Kaiser, H. M. and Blake, R. W., "Modeling Parasite Control Programs for Developing Nations Using Goal Programming," *Journal of Production Agriculture*, Vol. 4, No. 1 (January-March 1991), pp. 33-38.

Johnson, R. R., Zorn, Thomas S. and Schniederjans, Marc J., "The Fallacy of the Interior Decorator Fallacy: How to Customize Portfolios," *The Mid-Atlantic Journal of Business*, Vol. 25, No. 5 (March 1989), pp. 41-52.

Joiner, Carl, "Academic Planning Through the Goal Programming Model," *Interfaces*, Vol. 10, No. 4 (August 1980), pp. 86-91.

Joiner, Carl and Drake, Albert E., "Governmental Planning and Budgeting with Multiple Objective Models," *Omega*, Vol. 11, No. 1 (1983), pp. 57-66.

Joiner, Carl, "Reverse Resource Allocation: A Multi-Model, Multi-Goal Approach," *Journal of Education Finance*, Vol. 7, No. 2 (Fall 1981), pp. 205-218.

Joksh, H. C., "Programming with Fractional Linear Objective Functions," *Navel Research Logistics Quaterly*, Vol. 11 (1964), pp. 197-204.

Jones, Lawrence and Kwak, N. K., "A Goal Programming Model for Allocating Human Resources for the Good Laboratory Practice Regulations," *Decision Sciences*, Vol. 13, No. 1 (January 1982), pp. 156-166.

Jones, R. G., "Analyzing Initial and Growth Financing for Small Business," *Management Accounting*, Vol. 61 (1979), pp. 30-38.

Jose, Virginia D. and Tabucanon, M. T., "Multiobjective Model for Selection of Priority Areas and Industrial Projects for Investment Promotion," *Engineering Costs and Production Economics*, Vol. 10, No. 2 (1986), pp. 173-184.

Kahalas, H. and Groves, D. L., "Modeling for Organizational Decision-Making: Profit vs. Social Values in Resource Management," *Journal of Environmental Management*, Vol. 6 (1978), pp. 73-84.

Kahalas, H. and Key, R., "A Decisionally Oriented Manpower Model for Minority Group Business," *Quaterly Review of Economics and Business*, Vol. 14 (1974), pp. 71-84.

Kahalas, H. and Satterwhite, W., "Variable Impact of an Electric Utility's Social Responsibility," *Journal of Environmental Management*, Vol. 6 (1978), pp. 9-26.

Kalro, A.H., Chaturvedi, G. and Sengupta, S., "A Note on Goal Programming Approach to a Type of Quality Control Problem/Reply," *Journal of the Operational Research Society*, Vol. 34, No. 5 (March 1983), pp. 437-440.

Kambo, N. S., Handa, B. R. and Bose, R. K., "A Linear Goal Prgramming Model for Urban Energy-Economy-Environment Interaction," *Energy and Buildings*, Vol. 16, Nos. 1-2 (1991), pp. 537-551.

Kananen, I., Korhonen, P., Wallenius, J. and Wallenius, H., "Multiple Objective Analysis of Input - Output Models for Emergency Management," *Operations Reseach*, Vol. 38, No. 2 (1990), pp. 193-201.

Kang, Bong-Soon, "A Linear Goal Programming Model for Farm Planning of Semi-Subsistence Farms," *Journal of Rural Development* (Korea), Vol. 6, No. 2 (1983), pp. 87-105.

Kao, C. and Brodie, J. D., "Goal Programming for Reconciling Economic, Even-Flow, and Regulation Objectives in Forest Harvest Scheduling," *Canadian Joural of Forest Research*, Vol. 9, No. 4 (1979), pp. 525-531.

Karandikar, H., M. and Farrokh, Mistree, "Conditional Post-Solution Analysis of Multiobjective Compromise Decision Support Problem," *Engineering Optimization*, Vol. 12 (1987), pp. 43-61.

Karmarkar, U. S., "Convex/Stochastic Programming and Multilocation Inventory Problems," *Naval Research Logistics Quarterly*, Vol. 26, No. 1 (1979), pp. 1-19.

Kalu, T. C. U., "Determining the Impact of Nigeria's Economic Crisis on the Multimational Oil Companies: A Goal Programming Approach," *Journal of the Operational Research Society*, Vol. 45, No. 2 (1994), pp. 165-177.

Kendall, Kenneth E. and Lee, Sang M., "Formulating Blood Rotation Policies with Multiple Objectives," *Management Science*, Vol. 26, No. 11 (November 1980a), pp. 1145-1157.

Kendall, Kenneth E. and Lee, Sang M., "Improving Perishable Product Inventory Management Using Goal Programming," *Journal of Operations Management*, Vol. 1, No. 2 (1980b), pp. 77-84.

Kendall, Kenneth E. and Luebbe, R. L., "Management of College Recruiting Activities Using Goal Programming," *Decision Sciences*, Vol. 12, No. 2 (1981), pp. 193-205.

Kendall, Kenneth E., "Multiple Objective Planning for Regional Blood Centers," *Long Range Planning*, Vol. 13, No. 4 (1980), pp. 90-104.

Kendall, Kenneth E. and Schniederjans, Marc J., "Multi-Product Production Planning: A Goal Programming Approach," *European Journal of Operational Research*, Vol. 20, No. 1 (1985), pp. 83-91.

Kennedy, Peter, "Teaching Tips: Comparing Classification Techniques," *International Journal of Forecasting*, Vol. 7, No. 3 (November 1991), pp. 403-406.

Keown, Arthur J., "A Chance-Constrained Goal Programming Model for Bank Liquidity Management," *Decision Sciences*, Vol. 9, No. 1 (January 1978), pp. 93-106.

Keown, Arthur J. and Duncan, C. P., "Integer Goal Programming in Advertising Media Selection," *Decision Sciences*, Vol. 10, No. 4 (1979), pp. 577-592.

Keown, Arthur J. and Martin, J. D., "A Chance-Constrained Goal Programming Model for Working Capital Management," *The Engineering Economist*, Vol. 22, No. 3 (Spring 1977), pp. 153-174.

Keown, Arthur J. and Martin, J. D., "An Integer Goal Programming Model for Capital Budgeting in Hospitals," *Financial Management*, Vol. 5, No. 3 (Autumn 1976), pp. 28-35.

Keown, Arthur J. and Martin, J. D., "Capital Budgeting in the Public Sector: A Zero-One Goal Programming Approach," *Financial Management*, Vol. 7, No. 2 (Summer 1978), pp. 21-27.

Keown, Arthur J. and Taylor, B. W. III, "A Chance-Constrained Integer Goal Programming Model for Capital Budgeting in the Production Area," *Journal of the Operational Research Society*, Vol. 31, No. 7 (July 1980), pp. 579-589.

Keown, Arthur J. and Taylor, B. W. III, "Integer Goal Programming Model for the Implementation of Multiple Corporate Objectives," *Journal of Business Research*, Vol. 6, No. 3 (1978), pp. 221-235.

Keown, Arthur J., Taylor, B. W. III and Pinkerton, J. M., "Multiple Objective Capital Budgeting within the University," *Computers and Operations Reserach*, Vol. 8, No. 2 (1981), pp. 59-70.

Keown, Arthur J., Taylor, B. W. III and Duncan, C. P., "Allocation of Research and Development Funds: A Zero-One Goal Programming Approach," *Omega*, Vol. 7, No. 4 (1979), pp. 345-351.

Khorramshahgol, Reza and Azani, H., "A Decision Support System for Effective Systems Analysis and Planning," *Journal of Information and Optimization Sciences*, Vol. 9 (1988), pp. 41-52.

Khorramshahgol, Reza and Gousty, Yvon, "Delphic Goal Programming (DGP) - A Multi-Objective Cost-Benefit Approach to R-and-D Potrfolio Analysis," *IEEE Transactions on Engineering Management*, Vol. 33, No. 3 (1986), pp. 172-175.

Khorramshahgol, Reza and Hooshiari, A., "Three Shortcomings of Goal Programming and Their Solutions," Journal of *Information and Optimization Sciences*, Vol. 12, No. 3 (September 1991), pp. 459-466.

Khorramshahgol, Reza and Ignizio, J. P., "Single and Multiple Decision-Making in Multiple Objective Environment," *Advances in Management Studies*, Vol. 3, Nos. 3-4 (1984), pp. 181-192.

Khorramshahgol, Reza and Steiner, H. M., "Resource Analysis in Project Evaluation: A Multicriteria Approach," *Journal of the Operational Research Society*, Vol. 39 (1988), pp. 795-803.

Khorramshahgol, Reza, Azani, H. and Gousty, Y., "Integrated Approach to Project Evaluation and Selection," *IEEE Transactions on Engineering Management*, Vol. 35, No. 4 (1988), pp. 265-270.

Khorramshahgol, Reza, Dadzie, K. Q. and Akaah, I. P., "Delphic Goal Programming as a Planning Tool in LDC's: An Application," *Journal of Information and Optimization Sciences*, Vol. 49 (1988), pp. 33-40.

Killough, L. N. and Souders, T. L., "A Goal Programming Model for Public Accounting Firms," *The Accounting Review*, Vol. XLVIII, No. 2 (1973), pp. 268-279.

Kim, C. Y., "A Manufacturer's Retail Store Selection Model in the Normative Distribution Channel Design: A Goal Programming Approach," *Akron Business and Economic Review*, Vol. 14, No. 3 (Fall 1983), pp. 41-47.

Kim, Gyu C. and Schniederjans, Marc J., "A Mulitple Objective Model for a Just-In-Time Manufacturing System Environment," *International Journal of Operations and Production Management*, Vol. 13, No. 12 (1993), pp. 47-61.

Kim, Eyong B. and Kim, Sung, C., "Design and Development of an Effective Teaching System for Goal Programming," *Interfaces*, Vol. 14, No. 3 (1992), pp. 30-33.

Kim, Fung L. and Eng, Ung Choo, "A Linear Goal Programming Model for Classification with Non-monotone Attributes," *Computers and Operations Reserach*, Vol. 20, No. 4 (1993), pp. 403-408.

King, A. S., "A Programming Procedure for Evaluating Personnel Policies," *Personnel Administrator*, Vol. 27, No. 9 (September 1982), pp. 82-95.

Kinory, S., "Goal Programming and Managerial Decision Making," *Management International Review*, Vol. 18, No. 2 (1978), pp. 101-109.

Klimberg, R., Revelle, C. and J. Cohon, "A Multiobjective Approach to Evaluating and Planning the Allocation of Inspection Resources," *European Journal of Operational Research*, Vol. 52, No. 1 (1991), pp. 55-64.

Klock, D. R. and Lee, Sang M., "A Note on Decision Models for Insurers," *The Journal of Risk and Insurance*, Vol. 41, No. 3 (1974), pp. 537-543.

Kluyver, C. A., "On the Importance of Goal-Norming in Non-Preemptive Goal Programming," OPSEARCH, Vol. 16 (1979), pp. 88-97.

Knoll, A. L. and Engelberg, A., "Weighting Multiple Objectives-The Churchman-Ackoff Technique Revisited," *Computers and Operations Research*, Vol. 5 (1978), pp. 165-177.

Knutson, D. L., Marquis, L. M., Ricchiute, D. N. and Saunders, G. J., "A Goal Programming Model for Achieving Racial Balance in Public Schools," *Socio-Economic Planning Sciences*, Vol. 14, No. 3 (1980), pp. 109-116.

Koelling, C. Patrick and Bailey, James E., "A Multiple Criteria Decision Aid for Personnel Scheduling," *AIIE Transactions*, Vol. 16, No. 4 (1984), pp. 299-307.

Koizumi, A. and Inakazu, T., "A Multipurpose Optimization Model for Area-Wide Sewerage Systems," *Environment and Planning Analysis*, Vol. 21, No. 8 (August 1989), pp. 1015-1026.

Kopsidas, G. C., "A New Pareto Optimal Solution in a Lagrange Decomposable Multi-objective Optimization Problem," *Journal of the Operational Research Society*, Vol. 42, No. 5 (May 1991), pp. 401-441.

Korhonen, Antti, "A Dynamic Bank Portfolio Planning Model with Multiple Scenarios, Multiple Goals and Changing Priorities," *European Journal of Operational Research*, Vol. 30, No. 1 (1987), pp. 13-23.

Korhonen, Pekka and Laakso, Jukka, "Solving Generalized Goal Programming Problems Using a Visual Interactive Approach," *European Journal of Operational Research*, Vol. 26, No. 3 (1986), pp. 355-363.

Korhonen, Pekka and Soismaa, M. "A Multiple Criteria Model for Pricing Alcoholic Beverages," *European Journal of Operational Research*, Vol. 37 (1988) pp. 165-175.

Korhonen, Pekka and Wallenius, Jyrki, "Using Qualitative Data in Multiple Objective Linear Programming," *European Journal of Operational Research*, Vol. 48, No. 1 (1990) pp. 81-87.

Kornbluth, J. S. H., "Accounting in Multiple Objectives Linear Programming," *The Accounting Review*, Vol. 49 (April 1974), pp. 284-295.

Kornbluth, J. S. H., "A Survey of Goal Programming," *Omega*, Vol. 1, No. 2 (1973), pp. 193-205.

Kornbluth, J. S. H. and Steuer, R. E., "Goal Programming with Linear Fractional Criteria," *European Journal of Operational Research*, Vol. 8, No. 1 (1981a), pp. 58-65.

Kornbluth, J. S. H. and Steuer, R. E., "Multiple Objective Linear Fractional Programming," *Management Science*, Vol. 27, No. 9 (September 1981b), pp. 1024-1039.

Kornbluth, J. S. H. and Vinso, J. D., "Capital Structure and the Financing of the Multinational Approach: A Fractional Multiobjective Approach," *Journal of Financial and Quantitative Analysis*, Vol. 17, No. 2 (1982), pp. 147-178.

Kornbluth, J. S. H., "Engineering Design: Applications of Goal Programming and Multiple Objective Linear and Geometric Programming," *International Journal of Production Research*, Vol. 24, No. 4 (1986), pp. 945-953.

Kornbluth, J. S. H., "Multiple Objective Dynamic Programming with Forward Filtering," *Computers and Operations Research*, Vol. 13, No. 4 (1986), pp. 517-524.

Kornbluth, J. S. H., "Ratio Goals Manpower Planning Models," *INFOR*, Vol. 21, No. 2 (1983), pp. 151-154.

Kumar, P. C. and Philippatos, G. C., "Conflict Resolution in Investment Decisions: Implementation of Goal Programming Methodology for Dual-Purpose Funds," *Decision Sciences*, Vol. 10, No. 4 (1979), pp. 562-576.

Kumar, P. C., Philippatos, G. C. and Ezzell, J. R., "Goal Programming and the Selection of Portfolios by Dual-Purpose Funds," *The Jounal of Finance*, Vol. 33, No. 1 (1978), pp. 303-310.

Kumar, P. C., Singh, N. and Tewari, N. K., "A Nonlinear Goal Programming Model for Multistage, Multiobjective Decision Problems with Application to Grouping and Loading Problem in a Flexible Manufacturing System," *European Journal of Operational Research*, Vol. 53, No. 2 (1991), pp. 166-171.

Kumar, P. C., Singh, N. and Tewari, N. K., "A Nonlinear Goal Programming Model for The Loading Problem in Flexible Manufacturing System," *Engineering Optimization*, Vol. 12 (1987), pp. 315-323.

Kvanli, Alan H. and Buckley, J. J., "On the Use of U-Shaped Penalty Functions for Deriving a Satisfactory Financial Plan Utilizing Goal Programming," *Journal of Business Research*, Vol. 14, No. 1 (1986), pp. 1-18.

Kvanli, Alan H., "Financial Planning Using Goal Programming," *Omega*, Vol. 8 (1980), pp. 207-218.

Kwak, N. K., Allen, T. D. and Schniederjans, Marc J., "A Multilevel Salary Compensation Model Using Goal Programming," *R.A.I.R.O.-Operations Reseach*, Vol. 16, No. 1 (1982), pp. 21-31.

Kwak, N. K. and Diminnie, Carol B., "A Goal Programming Model for Allocating Operating Budgets of Academic Units," *Socio-Economic Planning Sciences*, Vol. 21, No. 5 (1987), pp. 333-339.

Kwak, N. K. and Jones, L., "An Application of PERT to R and D Scheduling," *Information Processing and Management*, Vol. 14 (1978), pp. 121-132.

Kwak, N. K. and Schniederjans, Marc J., "A Goal Programming Model as an Aid in Facility Location Analysis," *Computers and Operations Reserach*, Vol. 12, No. 2 (1985a), pp. 151-161.

Kwak, N. K. and Schniederjans, Marc J., "A Goal Programming Model for Improved Transportation Problems Solutions," *Omega*, Vol. 7, No. 4 (1979), pp. 367-370.

Kwak, N. K. and Schniederjans, Marc J., "A Goal Programming Model for Selecting a Facility Location Site," R.A.I.R.O.-*Operations Reseach*, Vol. 19, No. 1 (1985b), pp. 1-15.

Kwak, N. K. and Schniederjans, Marc J., "Letter: An Alternative Method for Solving Goal Programming Problems: A Reply," *Journal of the Operational Research Society*, Vol. 33, No. 9 (September 1982), pp. 859-860.

Kwak, N. K. and Schniederjans, Marc J., "An Alternative Solution Method for Goal Programming Problems: The Lexicographic Goal Programming Case," *Socio-Economic Planning Sciences*, Vol. 19, No. 2 (1985c), pp. 101-107.

Kwak, N. K. and Schniederjans, Marc J., "Goal Programming Solutions to Transportation Problems with Variable Supply and Demand Requirements," *Socio-Economic Planning Sciences*, Vol. 19, No. 2 (1985d), pp. 95-100.

Kwak, N.K. Schniederjans, Marc J. and Warkentin, Kimberly, S., "An Application of Linear Goal Programming to the Marketing Distribution Decision," *European Journal of Operational Research*, Vol. 52, No. 3 (1991) pp. 334-344.

Lam, K. F., Choo, E. U. and Wedley, W. C., "Linear Goal Programming in Estimation of Classification Probabilities," *European Journal of Operational Research*, Vol. 67, No. 1 (1993), pp. 101-110.

Lam, K. F. and Choo, E. U., "A Linear Goal Programming Model for Classification with Non-Monotone Attributes," *Computers and Operations Reserach*, Vol. 20, No. 4 (1993), pp. 403-408.

Lara, P. and Romero, Carlos., "An Interactive Multigoal Programming Model for Determining Livestock Rations: An Application to Dairy Cows in Andalusia, Spain," *Journal of the Operational Research Society*, Vol. 43, No. 10 (1992), pp. 945-953.

Lashine, S., Foote, B. and Ravindran, A., "A Nonlinear Mixed Integer Goal Programming Model for the Two-Machine Closed Flow Shop," *European Journal of Operational Research*, Vol. 55, No. 1 (1991), pp. 57-70.

Laurent, Gilles, "A Note on Range Programming: Introducing a "Satisficing Range" in a L.P.," *Management Science*, Vol. 22, No. 6 (1976), pp. 713-716.

Lawrence, Kenneth. D. and Burbridge, J. J., "A Multiple Goal Linear Programming Model for Coordinated Production and Logistics Planning," *International Journal of Production Research*, Vol. 14, No. 2 (1976), pp. 215-222.

Lawrence, Kenneth D.and Reeves, Gary R., "A Zero-One Goal Programming Model for Capital Budgeting in a Property and Liability Insurance Company," *Computers and Operations Reserach*, Vol. 9, No. 4 (1982), pp. 303-309.

Lawrence, Kenneth D., Reeves, Gary R. and Lawrence, Sheila M., "A Multiple Objective Shift Allocation Model," *AIIE Transactions*, Vol. 16, No. 4 (1984), pp. 323-328.

Lawrence, Sheila M., Lawrence Kenneth D., and Reeves, Gary R., "Allocation of Teaching Personnel: A Goal Programming Model," *Socio-Economic Planning Sciences*, Vol. 17, No. 4 (1983), pp. 211-216.

Lee, Bum-il, Chung, Nam-Kee and Tcha, Dong-Wan, "A Parallel Algorithm and Duality for a Fuzzy Multi-Objective Linear Fractional Programming Problem," *Computers and Industrial Engineering*, Vol. 20, No. 3 (1991) pp. 367-372.

Lee, Sang M., "A Gradient Algorithm for Chance Constrained Nonlinear Goal Programming," *European Journal of Operational Research*, Vol. 22, No. 3 (1985a), pp. 359-369.

Lee, Sang M., "An Aggregative Resource Allocation Model for Hospital Administration," *Socio-Economic Planning Sciences*, Vol. 7, No. 4 (1973), pp. 381-395.

Lee, Sang M., "A Revised Iterative Algorithm for Decomposition Goal Programming," *International Journal of Systems Science*, Vol. 14, No. 12 (1983), pp. 1383-1393.

Lee, Sang M. and Bird, M. M., "A Goal Programming Model for Sales Effort Allocation," *Business Perspectives*, Vol. 6, No. 4 (1970), pp. 17-21.

Lee, Sang M. and Chesser, D. L., "Goal Programming for Portfolio Selection," *The Journal of Portfolio Management*, Vol. 6, No. 3 (Spring 1980), pp. 22-26.

Lee, Sang M. and Chesser, D. L., "Goal Programming, Consulting, and Changing Times," *Oil and Gas Tax Quarterly*, Vol. 38, No. 1 (1989), pp. 164-182.

Lee, Sang M. and Clayton, E. R., "A Goal Programming Model for Academic Resource Allocation," *Management Science*, Vol. 18, No. 8 (1972), pp. B395-B408.

Lee, Sang M. and Clayton, E. R., "A Mathematical Programming Model for Academic Planning," *Southern Journal of Business*, Vol. 5, No. 4 (1970a), pp. 117-127.

Lee, Sang M. and Clayton, E. R., "Applications of Goal Programming for the Textile Industry," *Review of Industrial Management and Textile Science*, Vol. 6 (1970b), pp. 107-113.

Lee, Sang M. and Clayton, E. R., "Goal Programming for Academic Resource Allocation," *Management Science*, Vol. 26, No. 12 (December 1980a), pp. 1289-1290.

Lee, Sang M. and Clayton, E. R., "Management Science Update--A Goal Programming Model for Academic Resource Allocation," *Management Science*, Vol. 26, No. 12 (December 1980b), pp. 1289-1290.

Lee, Sang M. and Eom, Hyun B., "A Multi-Criteria Approach to Formulating International Project-Financing Strategies," *Journal of the Operational Research Society*, Vol. 40, No. 6 (1989), pp. 519-528.

Lee, Sang M. and Eom, Hyun B., "Financial and Capital Structuring Strategies for the Multinational Corporation," *International Economic Review*, Vol. 2 (1984), pp. 269-297.

Lee, Sang M. and Eom, Hyun B., "Multiple Criteria Decision Support Systems: The Powerful Tool for Attacking Complex, Unstructured Decisions," *Systems Practice*, Vol. 3, No. 1 (1990), pp. 51-62.

Lee, Sang M. and Franz, L., "Optimizing the Location-Allocation Problem with Multiple Objectives," *International Journal of Physical Distribution and Materials Management*, Vol. 9, No. 6 (1979), pp. 245-255.

Lee, Sang M. and Hall, P., "Decision Analysis and Risk Analysis: Management Science Does Apply to Strategy and Policy," *Management Science and Policy Analysis*, Vol. 5, No. 3 (1988), pp. 16-23.

Lee, Sang M. and Jaaskelainen, V., "Goal Programming: Management's Math Model," *Industrial Engineering*, Vol. 3, No. 1 (1971), pp. 30-35.

Lee, Sang M. and Jung, Hun-Joo, "Multi-Objective Production Planning Model in a Flexible Manufacturing Environment," *International Journal of Production Research*, Vol. 27, No. 11 (November 1989), pp. 1981-1992.

Lee, Sang M. and Keown, A. J., "Integer Goal Programming Model for Urban Renewal Planning," *Urban Systems*, Vol. 4, No. 1 (1979), pp. 17-26.

Lee, Sang M. and Klock, D., "A Note on Decision Models for Insurers," *Journal of Risk and Insurance*, Vol. 61, No. 3 (September 1974), pp. 537-543.

Lee, Sang M. and Kim, E. B., "An Analysis of Pedagogical Effects of Alternative Teaching Approaches of Goal Programming," *Decision Sciences*, Vol. 23, No. 4 (1992a), pp. 991-1002.

Lee, Sang M. and Kim, E. B., "The Pedagogical Effectiveness of Current Teaching Practices of Goal Programming," *Interface: Computer Education Quarterly*, Vol. 14, No. 2 (1992b), pp. 36-39.

Lee, Sang M. and Lerro, A. J., "A Cash Management Model for Health Care Clinics," *The Financial Review*, (1974a), pp. 1-10.

Lee, Sang M. and Lerro, A. J., "Capital Budgeting for Multiple Objectives," *Financial Management*, Vol. 3, No. 1 (Spring 1974b), pp. 58-66.

Lee, Sang M. and Lerro, A. J., "Optimizing the Portfolio Selection for Mutual Funds," *The Journal of Finance*, Vol. 28, No. 5 (1973), pp. 1087-1101.

Lee, Sang M. and Litschert, R., "Multidimensional Analysis of Organizational Identification," *Organization and Administrative Science*, Vol. 6, No. 4 (Winter 1976), pp. 29-32.

Lee, Sang M. and Luebbe, Richard L., "The Multi-Criteria Warehouse Location Problem Revisited," *International Journal of Physical Distribution and Materials Management*, Vol. 17, No. 3 (1987a), pp. 56-59.

Lee, Sang M. and Luebbe, Richard L., "A Comparison of A Constraint Aggregation and Partitioning Zero-One Goal programming Algorithm With The Lee and Morris Algorithm," *Computers and Operations Reserach*, Vol. 15, No. 2 (1988), pp. 97-102.

Lee, Sang M. and Luebbe, Richard L., "A Zero-One Goal-Programming Algorithm Using Partitioning and Constraint Aggregation," *Journal of the Operational Research Society*, Vol. 38, No. 7 (1987b), pp. 633-641.

Lee, Sang M. and Moore, Laurence J., "A Practical Approach to Production Scheduling," *Production and Inventory Management*, Vol. 15, No. 1 (1974a), pp. 79-92.

Lee, Sang M. and Moore, Laurence J., "Multi-Criteria School Busing Models," *Management Science*, Vol. 23, No. 7 (1977), pp. 703-715.

Lee, Sang M. and Moore, Laurence J., "Optimizing Transportation Problems with Multiple Objectives," *AIIE Transactions*, Vol. 5, No. 4 (1973), pp. 333-338.

Lee, Sang M. and Moore, Laurence J., "Optimizing University Admission Planning," *Decision Sciences*, Vol. 5, No. 3 (1974b), pp. 405-414.

Lee, Sang M. and Moore, Laurence J., "Optimizing University Admission Planning: Reply to Comment," *Decision Sciences*, Vol. 6, No. 1 (1975), pp. 192-193.

Lee, Sang M. and Morris, R. L., "Integer Goal Programming Methods," TIMS Studies in the *Management Sciences*, Vol. 6 (1977), pp. 273-289.

Lee, Sang M. and Nicely, R. E., "Goal Programming for Marketing Decisions: A Case Study," *Journal of Marketing*, Vol. 38, No. 1 (January 1974), pp. 24-32.

Lee, Sang M. and Olson, David L., "A Gradient Algorithm for Chance- Constrained Nonlinear Goal Programming," *European Journal of Operational Research*, Vol. 22, No. 3 (1985), pp. 359-369.

Lee, Sang M. and Olson, David L., "A Multicriteria Model for Regional Economic Planning," *Journal of Technology Transfer*, Vol. 5, No. 2 (Spring 1981), pp. 43-60.

Lee, Sang M. and Olson, David L., "A Zero-One Goal Programming Approach to Multi-Project Scheduling," *Project Management Journal*, Vol. 15, No. 2 (1984), pp. 61-74.

Lee, Sang M. and Olson, David L., "Chance Constrained Aggregate Blending," *Journal of Construction Engineering and Management*, Vol. 109, No. 1 (1983), pp. 39-47.

Lee, Sang M. and Puelz, Amy von, "Structuring Tax-Exempt Serial Revenue Bonds: A Multiple-Objective Decision Support System Framework," *Municipal Finance Journal*, Vol. 10, No. 2 (1989), pp. 153-171.

Lee, Sang M. and Rho, B. H., "A Multicriteria Decomposition Model for Two-Level, Decentralized Organization," *International Journal of Policy and Information*, Vol. 9, No. 1 (1985), pp. pp. 119-134.

Lee, Sang M. and Rho, B. H., "Computational Experience with the Dantzig-Wolfe and Kornai-Liptak Decomposition Algorithm," *International Journal of Policy and Information*, Vol. 10, No. 1 (1986), pp. 83-94.

Lee, Sang M. and Rho, B. H., "The Binary Search Decomposition in a Decentralized Organization," *Theory and Decision*, Vol. 11, No. 4 (1979a), pp. 353-362.

Lee, Sang M. and Rho, B. H., "The Modified Kornia-Liptak Decomposition Algorithm," *Computers and Operations Reserach*, Vol. 6, No. 1 (1979b), pp. 39-45.

Lee, Sang M. and Schniederjans, Marc J., "A Multicriteria Assignment Problem: A Goal Programming Approach," *Interfaces*, Vol. 13, No. 4 (1983), pp. 75-81.

Lee, Sang M. and Sevebeck, W. R., "An Aggregative Model for Municipal Economic Planning," *Policy Sciences*, Vol. 2, No. 2 (1971), pp. 99-115.

Lee, Sang M. and Shim, Jung P., "A Note on Microcomputer Applications in Decision Science," *Interface: The Computer Education Quaterly*, Vol. 6, No. 3 (1984), pp. 22-27.

Lee, Sang M. and Shim, Jung P., "Dissertation Research on Goal Programming," *Omega*, Vol. 15, No. 5 (1987a), pp. 345-347.

Lee, Sang M. and Shim, Jung P., "Interactive Goal Programming on the Microcomputer to Establish Priorities for Small Business," *Journal of the Operational Research Society*, Vol. 37. No. 6 (June 1986), pp. 571-577.

Lee, Sang M. and Shim, Jung P., "Multiple Objective Decision Making on the Microcomputer for Production/Operations Management: An Overview," *Socio-Economic Planning Sciences*, Vol. 21, No. 1 (1987b) pp. 33-36.

Lee, Sang M. and Shim, Jung P., "Planning for Multiple Objectives with Zero-Based Budgeting," *Long Range Planning*, Vol. 17, No. 5 (1984), pp. 103-110.

Lee, Sang M. and Soyibo, Adedoyin, "A Multiobjective Planning Model for University Resource Allocation," *European Journal of Operational Research*, Vol. 27 (1986), pp. 168-178.

Lee, Sang M. and Van Horn, J. C., "Analysis of Resource Allocation and Academic Policies Through Goal Programming," *Asian Economies*, No. 27 (1978), pp. 5-25.

Lee, Sang M. and Wilkins, Susan J., "Computer Facility Centralization/ Decentralization: A Multiobjective Analysis Model," *Computers and Operations Reserach*, Vol. 10, No. 1 (1983), pp. 29-40.

Lee, Sang M., Chung, S. H. and Everett, A. M., "Goal Programming Methods for Implementation of Just-In-Time Production," *Production Planning and Control*, Vol. 3, No. 2 (1992), pp. 175-182.

Lee, Sang M., Clayton, E. R. and Taylor, B. W. III, "A Goal Programming Approach to Multi-Period Production Line Scheduling," *Computers and Operations Reserach*, Vol. 5, No. 3 (1978), pp. 205-211.

Lee, Sang M., "Decision Analysis Though Goal Programming," *Decision Sciences*, Vol. 2, No. 2 (1971a), pp. 172-180.

Lee, Sang M., "Decision Analysis Though Goal Programming: A Substitute Solution," *Decision Sciences*, Vol. 2, No. 3 (1971b), pp. 377-378.

Lee, Sang M., Franz, L. S. and Wynne, A. J., "Optimizing State Patrol Manpower Allocation," *Journal of the Operational Research Society*, Vol. 30, No. 10 (October 1979), pp. 885-896.

Lee, Sand M., Gen, M. and Rho, B. H., "A Revised Iterative Algorithm for Decomposition Goal Programming," *International Journal of Systems Science*, Vol. 14, No. 12 (1983), pp. 1383-1393.

Lee, Sang M., "Goal Programming for Decision Analysis of Multiple Objectives," *Sloan Management Review*, Vol. 14, No. 2 (Winter 1972-73), pp. 11-24.

Lee, Sang M., "Goal Programming: Management's Math Model," *Journal of Industrial Engineering*, (January 1971c), pp. 104-108.

Lee, Sang M., Green, G. I. and Kim, Chang S., "A Multiple Criteria Model for the Location-Allocation Problem," *Computers and Operations Reserach*, Vol. 8, No. 1 (1981), pp. 1-8.

Lee, Sang M., Justis, R. and Franz, L., "Goal Programming for Decision Making in Closely Held Businesses," *American Journal of Small Business*, Vol. 3, No. 4 (1979), pp. 31-41.

Lee, Sang M., "LEEGP: Program for Goal Programming," *Journal of Marketing Research*, Vol. 10 (May 1970), pp. 199-200.

Lee, Sang M., Lerro, A. J. and McGinnis, B. R., "Optimization of Tax Switching for Commercial Banks," *Journal of Money, Credit, and Banking*, Vol. 3, No. 2 (February 1971), pp. 293-303.

Lee, Sang M., Luthans, Fred and Olson, David L., "A *Management Science* Approach to Contingency Models of Organizational Structure," *Academy of Management Journal*, Vol. 25, No. 3 (1982), pp. 553-566.

Lee, Sang M., "Management by Multiple Objectives," *Ekonomski Glansnik*, Vol. 35, No. 4 (1985b), pp. 685-690.

Lee, Sang M., Park, O. E. and Economides, S. C., "Resource Planning for Multiple Projects," *Decision Sciences*, Vol. 9, No. 1 (1978), pp. 49-67.

Lee, Sang M., Shim, Jung P. and Lee, C. S., "The Signal Flow Graph Method of Goal Programming," *Computers and Operations Research*, Vol. 11, No. 3 (1984), pp. 253-265.

Lee, Sang M., Synder, C. and Brisch, H., "Demilitarized Zone for Multiple Objectives," *Mathematical Social Sciences*, Vol. 5 (1983), pp. 33-46.

Lee, Sang M.,Tang, H., Olson, D. and Yen, D., "Formulating Industrial Development Policies: A Zero-One Goal Programming Approach," *Information and Management Sciences*, Vol. 13, No. 2 (1989), pp. 77-100.

Leinbach, Thomas R. and Cromley, Robert G., "A Goal Programming Approach to Public Investment Decisions: A Case Study of Rural Roads in Indonesia," *Socio-Economic Planning Sciences*, Vol. 17, No. 1 (1983), pp. 1-10.

Levary, Reuven R. and Avery, M. L., "On Practical Application of Weighting Equities in a Portfolio via Goal Programming," *Opsearch*, Vol. 21, No. 4 (1984), pp. 246-261.

Levary, Reuven R. and Choi, Tae S., "A Linear Goal Programming Model for Planning the Exports of Emerging Countries," *Journal of the Operational Research Society*, Vol. 34, No. 11 (1983), pp. 1057-1067.

Levary, Reuven R., "Optimal Control Problem with Goal Objectives Functions," *International Journal of Systems Science*, Vol. 17, No. 1 (January 1986a), pp. 97-109.

Levary, Reuven R., "Optimal Control Problem with Multiple Goal Objectives," *Optimal Control Applications and Methods*, Vol. 7, No. 2 (1986b), pp. 201-207.

Lilly, C. C. and Gleason, J. M., "Implications of Goal Programming for Insurance Agency Decision Making," *Omega*, Vol. 4, No. 3 (1976), pp. 353-354.

Ling-Hwie, C., Kwo-Jean, F. and Ching-Shu, T., "A Distributed Data Allocation Algorithm Based on the Zero-One Goal Programming Model," *Journal of Chinese Institute of Engineers*, Vol. 14, No. 5 (1991), pp. 551-558.

Linke, C. M. and Whitford, D. T., "A Multiobjective Financial Planning Model Model for Electric Utility Rate Regulation," *Journal of Economics and Business*, Vol. 35, No. 3 (1983), pp. 313-330.

Lin, W. Thomas, "A Survey of Goal Programming Applications," *Omega*, Vol. 8, No. 1 (January 1980a), pp. 115-117.

Lin, W. Thomas, "An Accounting Control System Structured on Multiple Objective Planning Models," *Omega*, Vol. 8, No. 3 (1980b), pp. 375-382.

Lin, W. Thomas, "Multiple Objective Budgeting Models: A Simulation," *The Accounting Review*, Vol. LIII, No. 1 (1978), pp. 61-76.

Lin, W. Thomas, "Applications of Goal Programming in Accounting," *Journal of Business Finance and Accounting*, Vol. 6, No. 4 (Winter 1979), pp. 559-577.

Llena, J., "On Fuzzy Linear Programming," *European Journal of Operational Research*, Vol. 22, No. 2 (1985), pp. 216-223.

Lockett, A. G. and Muhlemann, A. P., "A Problem of Aggregate Scheduling: An Application of Goal Programming," *International Journal of Production Research*, Vol. 16, No. 2 (1978), pp. 127-135.

Loganathan, G. V and Bhattacharya, D., "Goal Programming Techniques for Optimal Reservoir Operations," *Journal of Resources Planning and Management*, Vol. 116, No. 6 (November/December 1990), pp. 820-838.

Lohani, B. N. and Adulbhan, P. "A Multiobjective Model for Regional Water Quality Management," *Water Resources Bulletin*, Vol. 15, No. 4 (1979), pp. 1028-1038.

Lonergan, S. C. and Cocklin, C., "The Use of Lexicographic Goal Programming In Economic/Ecological Conflict Analysis," *Socio-Economic Planning Sciences*, Vol. 22, No. 2 (1988), pp. 83-92.

Lootsma, F.A., Mensch, T.C.A. and F.A. Vos, "Multi-criteria Analysis and Budget Reallocation in Long-term Research Planning," *European Journal of Operational Research*, Vol. 47, No. 3 (1990) pp. 293-305.

Loucks, D. P., "An Application of Interactive Multiobjective Water Resources Planning," *Interfaces*, Vol. 8, No. 1 (1977), pp. 70-75.

Loucks, J. S. and Jacobs, F. R., "Tour Scheduling and Task Assignment of a Heterogeneous Work Force: A Heuristic Approach," *Decision Sciences*, Vol. 22, No. 4 (1991), pp. 719-738.

Ludwin, William G. and Chamberlain, P. A., "Habitat Management Decisions with Goal Programming," *Wildlife Society Bulletin*, Vol. 17, No. 1 (Spring 1989), pp. 20-23.

Mackulak, G. T, Moodie, C. L. and Williams, T. J., "Computerized Hierarchical Production Control in Steel Manufacturing," *International Journal of Production Research*, Vol. 18, No. 4 (1980), pp. 455-465.

Madey, Gregory R. and Dean, Burton V., "Strategic Planning for Investment in Rand Using Decision Analysis and Mathematical Programming," *IEEE Transactions on Engineering Management*, Vol. EM-32, No.2 (1985), pp. 84-90.

Mahajan, Jayashree and Valharia, Asoo J., "A Mutiobjective Approach and Empirical Application of Sales-organization Design," *Decision Sciences*, Vol. 21, No. 3 (Summer 1990).

Malakooti, B., "A Gradient-Based Approach for Solving Hierarchical Multi-Criteria Production Planning Problems," *Computers and Industrial Engineering*, Vol. 16, No. 3 (1989) pp. 407-418.

Malakooti, B., "An Interactive On-line Multi-Objective Optimization Approach with Application to Metal Cutting Turning Operation," *International Journal of Production Research*, Vol. 29, No. 3 (1991a), pp. 575-598.

Malakooti, B., "Multiple Criteria Decision Making Approach for the Assembly Line Balancing Problem," *International Journal of Production Research*, Vol. 29, No. 10 (October 1991b), pp. 1979-2001.

Markland, R. E. and Vickery, S. K., "The Efficient Computer Implementation of a Large-Scale Integer Goal Programming Model," *European Journal of Operational Research*, Vol. 26, No. 3 (1986), pp. 341-354.

Markowski, Carol A. and Ignizio, James P., "Duality and transformations in Multiphase and Sequential Linear Goal Programming," *Computers and Operations Research*, Vol. 10, No. 4 (1983a), pp. 321-333.

Markowski, Carol A. and Ignizio, James P., "Theory and Properties of the Lexicographic Linear Goal Programming Dual," *Large Scale Systems*, Vol. 5 (1983b), pp. 115-121.

Martel, A. and Price, W., "Stochastic Programming Applied to Human Resource Planning," *Journal of the Operational Research Society*, Vol. 32, No. 3 (March 1981), pp. 187-196.

Martel, J. M. and Aouni, B., "Incorporating the Decision Maker's Preferences in the Goal Programming Model," *Journal of the Operational Research Society*, Vol. 41, No. 12 (December 1990), pp. 1121-1132.

Marten, G. G. and Sancholuz, L. A., "Ecological Land-Use Planning and Carrying Capacity Evaluation in the Jalapa Region (Veracruz, Mexico)," *Agro-Ecosystems*, Vol. 8 (1982), pp. 83-124.

Martinez-Legaz, J. E., "Lexicographical Order and Duality in Multiobjective Programming," *European Journal of Operational Research*, Vol. 33, No. 3 (1988), pp. 342-348.

Masud, Adu S. M. and Hwang, C. L., "Interative Sequential Goal Programming," *Journal of the Operational Research Society*, Vol. 32, No. 5 (May 1981), pp. 391-400.

Masud, Adu S. M. and Hwang, C. L., "An Aggregate Production Planning Model and Application of Three Multiple Objective Decision Methods," *International Journal of Production Research*, Vol. 18, No. 6 (1980), pp. 741-752.

Masud, Abu S. M. and Zheng, Xitong, "An Algorithm for Multiple-Objective Non-linear Programming," *Journal of the Operational Research Society*, Vol. 40, No. 10 (1989) pp. 895-906.

McCammon, D. F. and Thompson, W. Jr., "The Design of Tinplitz Piezoelectric Transducers Using Goal Programming," *Journal of Acoustic Society of America*, Vol. 68 (1980), pp. 754-757.

McCammon, D. F. and Thompson, W. Jr., "The Use of Goal Programming in the Design of Electroacoustic Transducers and Transducer Arrays," *Computers and Operations Reserach*, Vol. 10, No. 4 (1983), pp. 345-355.

McCann-Rugg, Mary, White, Gregory P. and Endres, Jeanette M., "Using Goal Programming to Improve the Calculation of Diabetic Diets," *Computers and Operations Reserach*, Vol. 10, No. 4 (1983), pp. 365-373.

McCarl, B. A. and Blake, B. F., "Goal Programming via Multidimensional Scaling Applied to Senegalese Subsistence Farming: Reply," *American Journal of Agricultural Economics*, Vol. 65, No. 4 (1983), pp. 832-833.

McClure, Richard H. and Wells, Charles E., "Incorporating Sales Force Preferences in a Goal Programming Model for the Sales Resource Allocation Problem," *Decision Sciences*, Vol. 18, No. 4 (1987a), pp. 677-681.

McClure, Richard H. and Wells, Charles E., "Modeling Multiple Criteria in the Faculty Assignment Problem," *Socio-Economic Planning Sciences*, Vol. 21, No. 6 (1987b), pp. 389-394.

McGlone, T. A. and Calantone, R. J., "A Goal Programming Model for Effective Segment Determination: A Comment and Application," *Decision Sciences*, Vol. 23, No. 5 (1992), pp. 1231-1239.

McGregor, M. J. and Dent, J. B., "An Application of Lexicographic Goal Programming to Resolve the Allocation of Water from the Rakaia River (New Zealand)," *Agricultural Systems*, Vol. 41, No. 3 (1993), pp. 349-268.

McKillop, W. and Liu, G., "Modelling Disaggregated Lumber Demand and Supply by Constrained Estimation Techniques," *Candian Journal of Forest Research*, Vol. 20, No. 6 (June 1990), pp. 781-789.

McKnew, M. A. and Sauydan, C., and Coleman, B. J., "An Efficient Zero-One Formulation of the Multilevel Lot-Sizing Problem," *Decision Sciences*, Vol. 22 (1991), pp. 301-312.

Mehrez, A. and Ben-Arieh, D., "All-Unit Discounts, Multi-Item Inventory Model with Stochastic Demand, Service Level Constraints and Finite Horizon," *International Journal of Production Research*, Vol. 29, No. 8 (August 1991), pp. 1615-1628.

Mehta, A. J. and Rifai, A. K., "Application of Linear Programming vs. Goal Programming to Assignment Problem," *Akron Business and Economic Review*, Vol. 7, No. 4 (Winter 1976), pp. 52-55.

Mehta, A. J. and Rifai, A. K., "Goal Programming Application to Assignment Problems in Marketing," *Journal of Academy Marketing Science*, Vol. 7, No. 2 (1979), pp. 108-116.

Mellichamp, J. M., Dixon, W. L. Jr and Mitchell, S. L., "Ballistic Missile Defense Technology Management with Goal Programming," *Interfaces*, Vol. 10, No. 5 (1980), pp. 68-74.

Mendoza, G. A., "A Heuristic Programming Approach in Estimating Efficient Target Levels in Goal Programming," *Canadian Journal of Forest Research*, Vol. 16, No. 2 (1986), pp. 363-366.

Mendoza, G. A., Bare, B. B. and Campbell, G. E., "Multiobjective Programming for Generating Alternatives: A Multiple-Use Planning Example," *Forest Science*, Vol. 33 (June 1987), pp. 458-468.

Mendoza, G. A., "Goal Programming Formulations and Extensions: An Overview and Analysis," *Canadian Journal of Forest Research*, Vol. 17, No. 7 (1987), pp. 575-581.

Mersha, Tigineh, Meredith, Jack and McKinney, John, "A Grant Rationing Model for a Health Care System," *Socio-Economic Planning Sciences*, Vol. 21, No. 3 (1987), pp. 159-165.

Merville, L. J. and Petty, J. W., "Transfer Pricing for the Multinational Firm," *The Accounting Review*, Vol. LIII, No. 4 (1978), pp. 935-951.

Merville, L. J. and Tavis, L. A., "Long-Range Financial Planning," *Financial Management*, Vol. 3, No. 2 (Summer 1974), pp. 56-63.

Michalowski, Wojtek,, "Evaluation of a Multiple Criteria Iterative Programming Approach: An Experiment," *INFOR*, Vol. 25, No. 2 (1987), pp. 165-173.

Miller, D. M. and Davis, R. P., "A Dynamic Resource Allocation Model for a Machine Requirements Problem," *AIIE Transactions*, Vol. 10 (1978), pp. 237-243.

Min, Hokey and Storbeck, James, "On the Origin and Persistence of Misconceptions in Goal Programming," *Journal of the Operational Research Society*, Vol. 42, No. 1 (1991) pp. 301-312.

Min, Hokey, "A Multiobjective Retail Service Location Model for Fast Food Restaurants," *Omega*, Vol.15, No. 5 (1987a), pp. 429-441.

Min, Hokey, "A Multiple Objective Approach to Workforce Redeployment Planning," *Omega*, Vol. 18, No. 4 (1990) pp. 425-432.

Min, Hokey, "An Disaggregate Zero-One Goal Programming Model for the Flexible Staff Scheduling Problem," *Socio-Economic Planning Sciences*, Vol. 21, No. 4 (1987b), pp. 271-282.

Min, Hokey, "International Intemodal Choices via Chance-Constrained Goal Programming," *Transportation Research*, Part A (General), Vol. 25A, No. 6 (November 1991), pp. 351-362.

Min, Hokey, "The Dynamic Expansion and Relocation of Capacitated Public Facilities: A Multi-Objective Approach," *Computers and Operations Reserach*, Vol. 15, No. 3 (1988a), pp. 243-252.

Min, Hokey, "Three-Phase Hierachical Allocation of University Resources Via Interactive Fuzzy Goal Programming," *Socio-Economic Planning Sciences*, Vol. 22, No. 5 (1988b), pp. 229-239.

Minguez, Ines M., Romero, Carlos and Domingo, Joaquin, "Determining Optimum Fertilizer Combinations Through Goal Programming with Penalty Functions: An Application To Sugar Beet Production in Spain," *Journal of the Operational Research Society*, Vol. 39, No. 1 (1988), pp. 61-70.

Mitchell, B. R. and Bare, B. B., "A Separable Goal Programming Approach to Optimation Multivariate Sampling Designs for Forest Inventory," *Forest Science*, Vol. 27, No. 1 (1981), pp. 147-162.

Mitra, A. and Patankar, J.G., "A Multi-objective Model for Warranty Estimation," *European Journal of Operational Research*, Vol. 45, Nos. 2-3 (1990) pp. 347-355.

Mitra, A. and Patankar, J. G., "An Integrated Multicriteria Model for Warranty Cost Estimation and Production," *IEEE Transactions on Engineering Management*, Vol. 40, No. 3 (August 1993), pp. 300-311.

Mitra, A. and Patankar, J. G., "Warranty Cost Estimation: A Goal Programming Approach," *Decision Sciences*, Vol. 19, No. 2 (1988), pp. 409-423.

Miyaji, I., Ohno, K. and Mine, H., "Solution Method for Partioning Students into Groups," European Journal of Operational Research, Vol. 33, No. 1 (1988), pp. 82-90.

Miyajima, Masaru and Nakai, Masato, "The Municipal financial Planning Model: A Simultaneous Regression Equations and Goal Programming Approach," *European Journal of Operational Research*, Vol. 27, No. 2 (1986), pp. 158-167.

Mogharabi, S. N. and Ravindran, A., "Liquid-Waste Injection Planning via Goal Programming," *Computers and Industrial Engineering*, Vol. 22, No.4 (1992), pp. 423-434.

Mohandas, S. U., Phelps, T. A. and Ragsdell, K. M., "Structural Optimization Using a Fuzzy Goal Programming Approach," *Computers and Structures*, Vol. 37, No. 1 (1990), pp. 1-8.

Mohanty, R. P. and Chandran, R., "Goal Programming Applications for Some Problems in Material Management," *Engineering Costs and Production Economics*, Vol. 8, No. 3 (1984), pp. 157-164.

Mohanty, R. P. and Govindrajan, S., "Design of a Production Planning and Control System for a toolroom: A Case Study," *Engineering Costs and Production Economics*, Vol. 16, No. 2 (1989), pp. 81-90.

Mohanty, R. P. and Rathnakumar, N., "Multicriteria Location/Allocation Problem Analysis," *Journal of the Institutionof Engineers*, Vol. 65 (1984), pp. 1-5.

Mohanty, R. P. and Singh, R., "A Hierarchical Production Planning Approach for a Steel Manufacturing System," *International Journal of Operations and Production Management*, Vol. 12, No. 5 (1992), pp. 69-78.

Monarchi, D. E., Kisiel, C. C. and Duckstein, L., "Interactive Multiobjective Programming in Water Resources: A Case Study," *Water Resources Research*, Vol. 9, No. 4 (1973), pp. 837-850.

Moore, L. J., Taylor, B. W. III and Lee, Sang M., "Analysis of a Transshipment Problem with Multiple Conflicting Objectives," *Computers and Operations Reserach*, Vol. 5, No. 1 (1978), pp. 39-46.

Moore, L. J., Taylor, B. W. III, Clayton, E. R. and Lee, Sang M., "Analysis of a Multi-Criteria Project Crashing Model," *AIIE Transactions*, Vol. 10, No. 2 (1978), pp. 163-169.

Moores, B., Garrod, N. W. and Briggs, G. H., "The Student Nurse Allocation Problem: A Formulation," *Omega*, Vol. 6, No. 1 (January 1978), pp. 93-96.

Morey, Richard C., "Managing the Armed Services Delayed Entry Pools to Improve Productivity in Recruiting," *Interfaces*, Vol. 15, No. 5 (1985), pp. 81-90.

Morris, R. L. and Lerro, A. J., "A Comparison of Goal Programming and Multiobjective Linear Programming," *The Mid-Atlantic Journal of Business*, Vol. 23, No. 1 (Winter 1984/85), pp. 35-44.

Morse, J. N. and Clark, R., "Goal Programming in Transportation Planning: The Problem of Setting Weights," *Northeast Regional Science Review*, Vol. 5 (1975), pp. 140-147.

Muhlemann, Alan P. and Lockett, A. G., "Portfolio Modeling in Multiple-Criteria Situations Under Uncertainty: Rejoiner," *Decision Sciences*, Vol. 11, No. 1 (January 1980), pp. 178-180.

Muhlemann, Alan P., Lockett, A. G. and Gear, A. E, "Portfolio Modeling in Multiple-Criteria Situations Under Uncertainty," *Decision Sciences*, Vol. 9, No. 4 (1978), pp. 612-626.

Murphy, Catherine M. and Ignizio, James P., "A Methodology for Multicriteria Network Partitioning," *Computers and Operations Reserach*, Vol. 11, No. 1 (1984), pp. 1-11.

Musa, A. A. and Saxena, U., "Scheduling Nurses Using Goal-Programming Technique," *AIIE Transactions*, Vol. 16, No. 3 (1984), pp. 216-221.

Muthukude, P., Novak, J. L. and Jolly, C., "A Goal Programming Evalution of the Fisheries Development Plan for Sri Lankas Coastal Fishing Fleet, 1988-1991," *American Journal of Agricultural Economics*, Vol. 72, No. 5 (1990), pp. 26-39.

Nakayama, H. and Furukawa, K., "Satisfacing Trade-Off Method with an Application to Multiobjective Structural Design, *Large Scale Systems*, Vol. 8, No. 1 (1985), pp. 47-59.

Nanda, J., Kothari, D. P. and Lingamurthy, K. S., "Economic-Emission Load Dispatch through Goal Programming Techniques," *IEEE Transactions on Energy Conversion*, Vol. 3, No. 1 (1988), pp. 26-39.

Narasimhan, Ram, "Goal Programming in a Fuzzy Environment," *Decision Sciences*, Vol. 11, No. 2 (1980), pp. 325-336.

Narasimhan, Ram, "On Fuzzy Goal Programming--Some Comments," *Decision Sciences*, Vol. 12, No. 3 (1981), pp. 533-538.

Nayak, N. N., Basu, M. and Tripathy, P. K., "Optimal Solution of a Deterministic Transportation Problem by the Duality in Goal Progrmming Techniques Under a Preemptive Priority Structured Approach," *Optimization*, Vol. 20, No. 3 (1989), pp. 325-334.

Neal, Heather D., France J. and Treacher, T. T., "Using Goal Programming in Formulating Rations for Pregnant Ewes," *Animal Production*, Vol. 42, Part 1 (February 1986), pp. 97-104.

Neely, W. P., North, R. M. and Fortson, J. C., "An Operational Approach to Multiple Objective Decision Making for Public Water Resource Projects Using Integer Goal Programming," *American Journal of Agricultural Economics*, Vol. 59, No. 1 (1977), pp. 198-203.

Neely, W. P., North, R. M. and Fortson, J. C., "Planning and Selecting Multiobjective Projects by Goal Programming," *Water Resources Bulletin*, Vol. 12, No. 1 (1976), pp. 19-25.

Neely, W. P., Sellers, J. and North, R. M., "Goal Programming Priority Sensitivity Analysis: An Application in Natural Resources Decision Making Processes," *Interfaces*, Vol. 10, No. 5 (1980), pp. 83-88.

Nelson, Cynthia A. and Wolch, Jennifer R., "Intrametropolitan Planning for Community Based Residential Care: A Goal Programming Approach," *Socio-Economic Planning Sciences*, Vol. 19, No. 3 (1985), pp. 205- 212.

Nelson, M., "Application of Management Science to Water Resource Planning in Latin America," *European Journal of Operational Research*, Vol. 3, No. 1 (January 1979), pp. 285-302.

Newton, K., "Interpreting Goal Attainment in Chance-Constrained Goal Programming," *OMEGA*, Vol. 13, No. 1 (1985), pp. 75-78.

Ng, Kevin Y. K., "A Multicriteria Optimization Approach to Aircraft Loading," *Operations Reseach*, Vol. 40, No. 6 (November-December 1992), pp. 1200-1205.

Ng, Kevin Y. K., "Goal Programming Solutions for Heat and Mass Transfer Problems-A Feasibility Study," *Computers and Chemical Engineering*, Vol. 15, No. 8 (August 1991), pp. 539-547.

Ng, Kevin Y. K., "Goal Programming Method of Weighted Residuals and Optimal Control Problems," *IEEE Transactions on Systems, Man and Cybernetics*, Vol. smc-17, No.1 (1987), pp. 102-106.

Ng, Kevin Y. K., "Solution of Navier-Stokes Equations by Goal Programming," *Journal of Computational Physics*, Vol. 39 (1981), pp. 103-111.

Nijkamp, P. and Spronk, J., "Analysis of Production and Location Decision by Means of Multicriteria Analysis," *Engineering Process Enconomics*, Vol. 42, No. 3 (1979), pp. 285-302.

Nussbaum, D. A. "Goal Programming as an Aid to Resource Management," *Defense Systems Management Review*, Vol. 3, No. 2 (1980), pp. 28-33.

O'Grady, P.J. and Mennon, U., "A Concise Review of Flexible Manufacturing Systems and FMS Literature," *Computers in Industry*, Vol. 7, No. 2 (1986a), pp. 155-167.

O'Grady, P. J. and Mennon, U., "A Flexible Multiobjective Production Planning Framework for Automated Manufacturing Systems," *Engineering Costs and Production Economics*, Vol. 8, No. 3 (1984), pp. 189-198.

O'Grady, P.J. and Mennon, U., "A Multiple Criteria Approach for Production Planning of Automated Manufacturing," *Engineering Optimization*, Vol. 8 (1986b), pp. 161-175.

Odom, Pat R., Shannon, R. E. and Buckles, Billy P., "Multi-Goal Subset Selection Problems Under Uncertainty," *AIIE Transactions*, Vol. 11, No. 1 (1979), pp. 61-69.

Ogryczak, W., "A Counter Example to Transformations in Multiphase and Sequencial Linear Goal Programming," *Control and Cybernetics*, Vol. 17, No. 1 (1988a), pp. 79-84.

Ogryczak, W., "A Symmetric Duality Concept for Goal Linear Programming: Principal Results," *Control and Cygernetics,* Vol.15, Nos. 3-4 (1986), pp. 413-423.

Ogryczak, W., "Symmetric Duality Theory for Linear Goal Programming," *Optimization*, Vol. 19, No. 3 (1988b), pp. 373-396.

Olson David L., Venkataramanan, M., and Mote, John, "A Technique Using Analytical Hierachy Process in Multiobjective Planning Models," *Socio-Economic Planning Sciences*, Vol. 20, No. 6 (1986), pp. 361-368.

Olson, David L. and Swenseth, Scott R., "A Linear Approximation for Chance-Constrained Programming," *Journal of the Operational Research Society*, Vol. 38, No. 3 (1987), pp. 261-267.

Olson, David L., "Comparison of Four Goal Programming Algorithms," *Journal of the Operational Research Society*, Vol. 35, No. 4 (1984), pp. 347-354.

Olson, David L., "Review of Empirical Studies in Multiobjective Mathematical Programming: Subject Reflection of Nonlinear Utility and Learning," *Decision Sciences*, Vol. 23, No. 1 (January/February 1992), pp. 1-20.

Olve, N., "Budgeting Design and Organizational Capabilities: Multicrtierion Planning of Telephone Services," *Omega*, Vol. 9, No. 6 (1981), pp. 571-578.

Oral, M. and Kettani, O. "Modelling the Process of Multiattribute Choice," *Journal of the Operational Research Society*, Vol. 40, No. 3 (March 1989), pp. 281-291.

Orne, D. L., Rao, A. and Wallace, W. A., "Profit Maximization with the Aid of Goal Programming for Speculative Housing Estate Developers," *Operational Research Quaterly*, Vol. 26, No. 4 (1975), pp. 813-826.

Osinski, Z., Pokojski, J. and Wrobel, J., "Optimization of Multilevel Multicriteria Machine Design Problems," *Foundations of Control Engineering*, Vol. 8 (1983), pp. 175-182.

Osleeb, J. P. and Ratick, S. J., "A Mixed Integer and Multiple Objective Programming Model to Analyze Coal Handling in New England," *European Journal of Operational Research*, Vol. 12, No. 3 (March 1983), pp. 302-313.

Ozatalay, S and Broyles, R. W., "Optimum Distribution of Diagnostic-Specific Technology," *Journal of Medical Systems*, Vol. 8, No. 4 (1984), pp. 249-264.

Ozkarahan, Irem and Bailey, James E., "Goal Programming Model Subsystem of a Flexible Nurse Scheduling Support System," *AIIE Transactions*, Vol. 20, No. 3 (1988), pp. 306-316.

Ozkarahan, Irem, "Flexible Nurse Scheduling Support System," *Computer Methods and Programs in Biomedicine*, Vol. 30, Nos. 2-3 (1989), pp. 145-153.

Padmanabhan, G. and Vrat, Prem, "Analysis of Multi-Item Inventory Systems Under Resource Constraints: A NonLinear Goal Programming Approach," *Engineering Costs and Production Economics,* Vol. 20, No. 2 (1990), pp. 121-127.

Panagiotakopoulos, Demetros, "A Multi-Objective Framework for Environmental Management Using Goal Programming," *Journal of Environmental Systems,* Vol. 5, No. 2 (1975), pp. 133-147.

Pant, J. C. and Shah, S., "Linear Approach to Linear Fractional Goal Programming," *Opsearch,* Vol. 29, No. 4 (1992), pp. 297-304.

Papageorgio, J. C., "A Goal Programming Model for Resource Allocation in the School System of Greece," *Kybernetes,* Vol. 7 (1978), pp. 229-237.

Park, Yang B. and Koelling, C. Patrick, "A Solution of Vechile Routing Problems in a Multiple Objective Environment," *Engineering Costs and Production Economics,*Vol. 10, No. 2 (1986), pp.121-132.

Parker, Barnett R., "A Multiple Goal Programming Methodology for Evaluation Management Information Systems," *Omega,* Vol. 13 (1985), pp. 313-330.

Parker, Barnett R., "A Program Selection/Resource Allocation Model for Control of Malaria and Related Parasitic Diseases," Computer and *Operations Reseach,* Vol. 10, No. 4 (1983), pp. 375-389.

Parker, Barnett R. and Kaluzny, A. D., "Design Planning to Meet Goals in Human Service Organizations," Human Systems Management, Vol. 3, No. 2 (1982), pp. 77-90.

Parker, Barnett R., Mtango, F. D., Koda, G. R., Killewo, J. J., Muhondwa, E. P. and Newman, Jeanne S., "A Methodology for Design, Evaluation, and Improvement of Village Health Worker Supervision Schemes in Rural Tanzania," *Socio-Economic Planning Sciences,* Vol. 20, No. 4 (1986), pp. 219-232.

Patrick, G. F. and Blake, B. F., "Measurement and Modeling of Farmers' Goals: An Evaluation and Suggestions," *Southern Journal of Agricultural Economics,* Vol. 12, No. 1 (1980), pp. 199-204.

Pentzaropoulos, G. C. and Giokas, D. I., "Cost-Performance Modelling and Optimization of Network Flow Balance via Goal Programming Analysis," *Computer Communications*, Vol. 16, No. 10 (1993), pp. 645-651.

Perlis, J. H. and Ignizio, J. P., "Stability Analysis: An Approach to the Evalution of System Design," *Cybernetics and Systems: An International Journal*, Vol. 11 (June-August 1980), pp. 87-103.

Philippatos, G. C. and Christofi, A., "Liquid-Asset Management Modeling for Inter-Subsidary Operations of Multinational Corporations: A Goal Programming Approach," *Management International Review*, Vol. 24, No. 2 (1984), pp. 4-14.

Philipson, R. H. and Ravindran, A., "Application of Goal Programming to Machinability Data Optimization," *Journal of Mechanical Design*, Vol. 100, No. 2 (1978), pp. 286-291.

Philipson, R. H. and Ravindran, A., "Application of Mathematical Programming to Metal Cutting,"*Mathematical Programming Study*, Vol. 11 (1979), pp. 116-134.

Phillips, Nancy V., "A Weighting Function for Pre-Emptive Multi-Criteria Assignment Problems," *Journal of the Operational Research Society*, Vol. 38, No. 9 (1987), pp. 797-802.

Pickens, J. B. and Hof, John G., "Fuzzy Goal Programming in Forestry: An Application with Special Solution Problems," *Fuzzy Sets and Systems*, Vol. 39, No. 3 (1991), pp. 239-246.

Piech, B. and Rehman, T., "Application of Multiple Criteria Decision Making Methods to Farm Plannings: A Case Study," *Agricultural Systems*, Vol. 41, No. 3 (1993), pp. 305-319.

Piekanen, E., "Goal Programming and Operational Objectives in Public Administration," *Swedish Journal of Economics*, Vol. 72, No. 3 (1970), pp. 207-214.

Porterfield, R. L., "A Goal Programming Model to Guide and Evaluate Tree Improvement Programs," *Forest Science*, Vol. 22, No. 4 (1976), pp. 417-430.

Prakash, Jai, Sinha, S. B. and Sahay, S. S., "Bus Transportation Crews Planning by Goal Programming," *Socio-Economic Planning Sciences*, Vol. 18, No. 3 (1984), pp. 207-210.

Premachandra, I. M., "A Goal Programming Model for Activity Crashing in Project Networks," *International Journal of Operations and Production Management*, Vol. 13, No. 6 (1993), pp. 79-85.

Price, W. L. and Gravel, M., "Solving Network Manpower Problems with Side Constraints," *European Journal of Operational Research*, Vol. 15, No. 2 (February 1984), pp. 196-202.

Price, W. L. and Piskor, W. G., "The Application of Goal Programming to Manpower Planning," *INFOR*, Vol. 10, No. 3 (1972), pp. 221-231.

Price, W. L., "A Review of Mathematical Models in Human Resources Planning," *Omega*, Vol. 8, No. 6 (1980), pp. 639-645.

Price, W. L., "Solving Goal Programming Manpower Models Using Advanced Network Codes," *Journal of the Operational Research Society*, Vol. 29, No. 12 (December 1978), pp. 1231-1239.

Proll, L. G., "Comments on the Alternative Solution Method for Goal Programming," *Journal of the Operational Research Society*, Vol. 33, No. 8 (August 1982), pp. 766-767.

Psarras, J., P. Capros, and J.-E. Samouilidis, "Multi Criteria Analysis Using a Large-Scale Energy Supply LP Model," *European Journal of Operational Research*, Vol.44, No. 2 (1990) pp. 175-184.

Puelz, Amy von and Lee, Sang M., "A Multiple-Objective Programming Technique for Structuring Tax-Exempt Serial Revenue Debt Issues," *Management Science*, Vol. 38, No. 8 (August 1992), pp. 1186-1200.

Qassim, R. Y. and Silveira, C. S., "Heat Exchanger Network Synthesis: The Goal Programming," *Computers and Chemical Engineering*, Vol. 12, No. 11 (November 1988), pp. 1163-1165.

Rae, A. N., "A Note on the Solution of Goal Programming Problems with Preemptive Priority," *New Zealand Operational Research*, Vol. 2 (1974), pp. 34-39.

Rakes, Terry R. and Franz, Lori S., "Decision Support Models for Project Planning," *Omega*, Vol. 13 (1985), pp. 73-74.

Rakes, Terry R., Franz, Lori S. and Wynne, A. James, "Aggregate Production Planning Using Chance-Constrainted Goal Programming," *International Journal of Production Research*, Vol. 22, No. 4 (1984), pp. 673-684.

Rao, A., "Quantity Discounts in Today's Market," *Journal of Marketing*, Vol. 44 (1980), pp. 44-51.

Rao, J. R., Tiwari, R. N. and Chakraborty, D., "Chance-Contrained Fuzzy Goal Programming," *Journal of Fuzzy Mathematics*, Vol. 1, No. 4 (1993), pp. 823-834.

Rao, J. R., Tiwari, R. N. and Mohanty, B. K., "A Preference Structure on Aspiration Levels in a Goal Programming Problem--A Fuzzy Approach," *Fuzzy Sets and Systems*, Vol. 25, No. 2 (February 1988a), pp. 175-182.

Rao, J. R., Tiwari, R. N. and Mohanty, B. K., "Preference Structure on Alterntives and Judges in a Group Decision Problem--A Fuzzy Approach," *International Journal of Systems Science*, Vol. 19, No. 9 (September 1988b), pp. 1795-1812.

Rao, S. S., Venkayya, V. B. and Knot, N. S., "Optimization of Activitiy Controlled Structures Using Goal Programming," *International Journal of Numerical Methods in Engineering*, Vol. 26, No. 1 (January 1988), pp. 183-197.

Rasmussen, L. M., "Zero-One Programming with Multiple Criteria," *European Journal of Operational Research*, Vol. 26, No. 1 (1986), pp. 83-95.

Ratick, S. J., "Multiobjective Programming with Related Bargaining Games," *Regional Science and Urban Economics*, Vol. 13, No. 1 (1983), pp. 141-160.

Ravindran, A., Shin, Wan S., Arthur, Jefferey L. and Moskowitz, Herbert, "Nonlinear Integer Goal Programming Models for Acceptance Sampling," *Computers and Operations Reserach*, Vol. 13, No. 5 (1986), pp. 611-622.

Ray, S. C., "Goal Programming - A Possible Approach for Decision Making in Mining Industry," *Journal of Mines, Metals and Fuels*, Vol. 34, Nos. 5-6 (May-June 1986), pp. 290-292.

Rees, Loren P., Clayton, Edward R. and Taylor, Bernard W. III, "Solving Multiple Response Simulation Models using Response Surface Methodology within a Goal Programming Framework," *AIIE Transactions*, Vol. 17, No. 1 (1985), pp. 47-57.

Reeves, G. R. and Lawrence, K. D., "Combining Multiple Forecasts Give Multiple Objectives," *Journal of Forecasting*, Vol. 1, No. 3 (1982), pp. 271-279.

Reeves, G. R. and Hedin, S. R., "A Generalized Interactive Goal Programming Procedure," *Computers and Operations Reserach*, Vol. 20, No. 7 (1993), pp. 747-753.

Reeves, G. R., "A Note on Quadratic Preferences and Goal Programming," *Decision Sciences*, Vol. 9, No. 3 (1978), pp. 532-534.

Rehman, Tahir and Romero, Carlos, "Goal Programming with Penalty Functions and Livestock Ration Formulation," *Agricultural Systems*, Vol. 23 (1987), pp. 117-132.

Rehman, Tahir and Romero, Carlos, "Multiple-Criteria Decision Making Techniques and Their Role in Livestock Ration Formulation," *Agricultural Systems*, Vol. 15 (1984), pp. 23-49.

Rehman, Tahir and Romero, Carlos, "The Application of the MCDM Paradigm to the Management of Agricultural Systems: Some Basic Considerations," *Agricultural Systems*, Vol. 41, No. 3 (1993), pp. 239-255.

Rensi, G. and Hrubes, R. J., "Implications of Goal Programming Forest Resource Allocation: A Rejoinder," *Forest Science*, Vol. 29, No. 4 (1983), pp. 841-842.

Retzlaff-Roberts, D. L. and Morey, R. C., "A Goal Programming Method of Stochastic Allocative Data Envelopment Analysis," *European Journal of Operational Research*, Vol. 71, No. 3 (1993), pp. 379-397

Reznicek, K. K., Simonovic, S. P. and Bector, C. R., "Optimization of Short-Term Operation of a Single Multipurpose Reservior--A Goal Programming Approach," *Canadian Journal of Civil Engineering*, Vol. 18, No. 3 (1991), pp. 397-406.

Rifai, Ahmed K. and Dey, M. K., "Goal Programming: An Effective Tool in Strategic Quality Planning," *Quality Forum*, Vol. 16, No. 4 (1990), pp. 198-203.

Rifai, Ahmed K. and Pecenka, Joseph O., "An Application of Goal Programming in Healthcare Planning," *International Journal of Operations and Production Management*, Vol. 10, No. 3 (1990), pp. 28-37.

Rifai, Ahmed K. and Pecenka, Joseph O., "Goal Achievement Through Goal Programming: Short Versus Long Term," *Engineering Costs and Production Economics*, Vol. 10, No. 2 (1986), pp. 155-160.

Rifai, Ahmed K., "Sensitivity Analysis of Linear Programming vs. Goal Programming," *Industrial Management*, Vol. 22, No. 1 (1980), pp. 12-17.

Rifai, Ahmed K., "The Role of Mathematical Programming in Allocating Scarce Resources: The Case of Multiple Goals," *Industrial Management*, Vol. 20, No. 4 (1978), pp. 1-3.

Ringuest, Jeffrey L. and Graves, S. B., "The Linear Multi-Objective R and D Project Selection Problem," *IEEE Transactions on Engineering Management*, Vol. 36, No. 1 (1989), pp. 54-56.

Ringuest, Jeffrey L. and Gulledge, Thomas R. Jr., "A Preemptive Value - Function Method Approach for Multiobjective Linear Programming Problems," *Decision Sciences*, Vol 14, No. 1 (1983), pp. 76-86.

Ritzman, Larry and Krajewski, L. J., "Multiple Objectives in Linear Programming - An Example in Scheduling Postal Resources," *Decision Sciences*, Vol. 4, No. 2 (1973), pp. 364-378.

Ritzman, Larry, Bradford, J. and Jacobs, R., "A Multiple Objective Approach to Space Planning for Academic Facilities," *Management Science*, Vol. 25, No. 9 (September 1979), pp. 895-906.

Rivett, P., "Multidimensional Scaling for Multiobjective Policies," *Omega*, Vol. 5, No. 4 (1977), pp. 367-379.

Ro, In-Kyo and Kim, Joong-In, "Multi-criteria Operational Control Rules in Flexible Manufacturing Systems," *International Journal of Production Research*, Vol. 28, No. 1 (1990) pp. 47-64.

Romero, Carlos, "A Note: Effects of Five-Sided Penalty Functions in Goal Programming," *Omega*, Vol. 12 (1984), p. 333.

Romero, Carlos, "A Survey of Generalized Goal Programming (1970-1982)," *European Journal of Operational Research*, Vol. 25, No. 2 (1986), pp. 183-191.

Romero, Carlos and Amador, Francisco, "A Note: Effects of Logarithmic Transformations in Nonlinear Goals in the Goal Programming Problem," *Engineering Optimization*,Vol. 9, No.4 (1986), pp.299-302.

Romero, Carlos and Rehman, Tahir, "Goal Programming and Multiple Criteria Decision Making in Farm Planning: An Expository Analysis," *Journal of Agricultural Economics*, Vol. 35, No. 2 (1984a), pp. 177-190.

Romero, Carlos and Rehman, Tahir, "A Note on Diet Planning in the Third World by Linear and Goal Programming," *Journal of the Operational Research Society*, Vol. 35, No. 6 (1984b), pp. 555-558.

Romero, Carlos and Rehman, Tahir, "Goal Programming and Multiple Criteria Decision-Making in Farm Planning: Some Extensions," *Journal of Agricultural Economics*, Vol. 35, No. 2 (1985), pp. 171-185.

Romero, Carlos and Rehman, Tahir, "Natural Resource Management and the Use of Multiple Criteria Decision-Making Techniques: A Review," *European Review of Agricultural Economics*, Vol. 14, No. 1 (1987), pp. 61-90.

Romero, Carlos and Rehman, Tahir, "Goal Programming via Multidimensional Scaling Applied to Senegalese Subsistence Farming: Comment," *American Journal of Agricultural Economics*, Vol. 65, No. (1983), pp. 829-831.

Romero, Carlos, "Multi-Objective and Goal-Programming Approaches as a Distance Function Model," *Journal of the Operational Research Society*, Vol. 36, No. 3 (1985a), pp. 249-251.

Romero, Carlos, "Note--Naive Weighting in Non-preemptive Goal Programming," *Journal of the Operational Research Society*, Vol. 36 (1985b), pp. 647-648.

Rosenblatt, M. J., "The Facilities Layout Problem: A Multi-Goal Approach," *International Journal of Production Research*, Vol. 17, No. 4 (1979), pp. 323-332.

Rosenbloom, E. S. and Shiu, Elias S. W., "The Matching of Assets with Liabilities by Goal Programming," *Managerial Finance*, Vol. 16, No. 1 (1990), pp. 23-31.

Rosenthal, R. E., "Goal Programming--A Critique," *New Zealand Operational Research*, Vol. 11 (1983), pp. 1-7.

Rosenthal, R. E., "Principles of Multiobjective Optimization," *Decision Sciences*, Vol. 16, No. 2 (1985), pp. 133-152.

Rowe, M. D. and Pierce, B. L., "Some Tests of Analytical Multiobjective Decision-Making Methods," *Socio-Economic Planning Sciences*, Vol. 16, No. 3 (1982), pp. 133-140.

Roy, B., "Problems and Methods with Multiple Objective Functions," *Mathematical Programming*, Vol. 1, No. 2 (1971), pp. 239-266.

Rubin, P. A. and Narasimhan, Ram, "Fuzzy Goal Programming with Nester Priorities," *Fuzzy Sets and Systems*, Vol. 14, No. 2 (1984), pp. 115-129.

Ruefli, Timothy W., "A Generalized Goal Decomposition Model," *Management Science*, Vol. 17, No. 8 (1971), pp. B505-B518.

Ruefli, Timothy W. and Storbeck, James, "A Model for Resource Allocation in Complex Hierarchies," *Socio-Econ Planning Sciences*, Vol. 18, No. 1 (1984), pp. 59-67.

Rumpf, D. L., "Estimating Postsecondary Student Flow with Limited Data," *Research in Higher Education*, Vol. 27, No. 1 (1987), pp. 39-50.

Rustangi, K. P. and Bare, B. B., "Resolving Multiple Goal Conflicts with Interactive Goal Programming," *Canadian Journal of Forest Research*, Vol. 17, No. II (November 1987), pp. 1401-1407.

Saatcioglu, Omer, "A Multi-Attribute Assignment Goal-Programming Model with Incentives," *Journal of the Operational Research Society*, Vol. 38, No. 4 (1987), pp. 361-365.

Saber, H. M. and Ravindran, A., "A Partitioning Gradient Based (PGB) Algorithm for Solving Nonlinear Goal Programming Problems," *Computers and Industrial Engineering*, Vol. 23, Nos. 1-4 (November 1992), pp. 291-294.

Saber, H. M. and Ravindran, A., "Nonlinear Goal Programming Theory and Practice: A Survey," *Computers and Operations Reserach*, Vol. 20, No. 3 (1993), pp. 275-292.

Sakawa, M., "Interactive Fuzzy Goal Programming for Multiobjective Nonlinear Programming Problems," *Electronics and Communication in Japan*, Part 1, Vol. 68, No. 11 (November 1985), pp. 49-56.

Saladin, B. A., "Goal Programming Applied to Police Patrol Allocation," *Journal of Operations Management*, Vol. 2, No. 4 (1982), pp. 239-249.

Salvia, A. A. and Ludwing, W. R., "An Application of Goal Programming at Lord Corporation," *Interfaces*, Vol. 9, No. 4 (1979), pp. 129-133.

Samouilidis, J. E. and Pappas, I. A., "A Goal Programming Approach to Energy Forecasting," *European Journal of Operational Research*, Vol. 5, No. 5 (1980), pp. 321-331.

Sandgren, Eric, "Structural Design Optimization for Latitude by Nonlinear Goal Programming," *Computers and Structures*, Vol. 33, No. 6 (1989), pp. 1395-1402.

Sandiford, Frances, "Analysis of Multiobjective Decision-Making for the Scottish Inshore-Fishery," *Journal of Agricultural Economics*, Vol. 37 (1986), pp. 207-219.

Sankaran, S., "Multiple Objective Decision Making Approach to Cell Formation: A Goal Programming Model," *Mathematical and Computer Modelling* (Oxford), Vol. 13, No. 9 (1990), pp. 71-82.

Santhanam, R., Muralidhar, K. and Schniederjans, Marc J., "A Zero-One Goal Programming Approach for Information System Project Selection," *Omega*, Vol. 17, No. 6 (1989) pp. 583-593.

Sardana, G. D. and Vrat, Prem, "Productivity Measurement in a Large Organization with Multi-Performance Objectives: A Case Study," *Engineering Management International*, Vol. 4, No. 2 (1987), pp. 105-125.

Sarma, G. V., Sellami, L. and Houam, K. D., "Application of Lexicographic Goal Programming in Production Planning--Two Case Studies," *Opsearch*, Vol. 30, No. 2 (1993), pp. 141-142.

Sartoris, W. L. and Spruill, M. L., "Goal Programming and Working Capital Management," *Financial Management*, Vol. 3, No. 1 (Spring 1974), pp. 67-74.

Sasaki, M. and Gen, M., "A Method for Solving Fuzzy Multibojective Decision Making Problems by Interactive Sequential Goal Programming," *Transactions of the Institute of Electronics, Information and Communication Engineers*, Vol. J75-A, No. 10 (1992), pp. 1590-1995.

Sasaki, M., Gen, M. and Ida, K., "Interative Sequential Fuzzy Goal Programming," *Computers and Industrial Engineering*, Vol. 19, Nos. 1-4 (1990), pp. 567-571.

Saunders, Gary, "An Application of Goal Programming to the Desegration Busing Problem," *Socio-Economic Planning Sciences*, Vol. 15, No. 6 (1981), pp. 291-293.

Schenkerman, Stan, "Use and Abuse of Weights in Multiple Objective Decision Support Models," *Decision Sciences*, Vol. 22, No. 2 (Spring 1991) pp. 369-378.

Schilling, D. A., Revelle, C. and Cohon, J., "An Approach to the Display and Analysis of Multiobjective Problems," *Socio-Economic Planning Sciences*, Vol. 17, No. 2 (1983), pp. 57-63.

Schinnar, A. P., "A Multidimensional Accounting Model for Demographic and Economic Planning Interactions," *Environment and Planning Analysis*, Vol. 8, No. 4 (1976), pp. 455-475.

Schniederjans, Marc J. and Fowler, K., "Strategic Acquisition Management: A Multi-objective Synergistic Approach," *Journal of the Operational Research Society*, Vol. 40, No. 4 (April 1989), pp. 333-345.

Schniederjans, Marc J. and Hoffman, J. J., "Multinational Acquistion Analysis: A Zero-One Goal Programming Model," *European Journal of Operational Research*, Vol. 62, No. 2 (1992), pp. 175-185.

Schniederjans, Marc J. and Kim, Gyu C., "A Goal Programming Model to Optimize Departmental Preferences in Course Assignments," *Computers and Operations Reserach*, Vol. 14, No. 2 (1987), pp.87-96.

Schniederjans, Marc J. and Kim, Sang O., "An Early Settlement of Long Term Strikes: A Game Theory and Goal Programming Approach," *Socio-Economic Planning Sciences*, Vol. 21, No. 3 (1987), pp.177-188.

Schniederjans, Marc J. and Kwak, N. K., "An Alternative Solution Method for Goal Programming Problems: A Tutorial," *Journal of the Operational Research Society*, Vol. 33, No. 3 (March 1982), pp. 247-251.

Schniederjans, Marc J. and Markland, Robert E., "Estimating Start-Up Resource Utilization in a Newly Formed Organization," *Interfaces*, Vol. 16, No. 5 (1986), pp. 101-109.

Schniederjans, Marc J. and Santhanam, R., "A Model Formulation System for Information System Project Selection," *Computers and Operations Reserach*, Vol. 20, No. 7 (September 1993a), pp. 755-767.

Schniederjans, Marc J. and Santhanam, R., "A Multi-Objective Constrained Resource Information System Project Selection Method," *European Journal of Operational Research*, Vol. 70, No. 2 (1993b), pp. 244-253.

Schniederjans, Marc J. and Santhanam, R., "A Zero-One Goal Programming Approach for Journal Selection and Cancellation Problem," *Computers and Operations Reserach*, Vol. 16, No. 6 (1989), pp. 557-565.

Schniederjans, Marc J. and Wilson, R.L., "Using the Analytic Hierarchy Process and Goal Programming for Information System Project Selection," *Information and Management*, Vol. 20, No. 5 (May 1991), pp. 333-342.

Schniederjans, Marc J., "A Note on Decision-Making Criteria for Algorithm Selection: Reducing Goal Programming Computational Effort," *Decision Sciences*, Vol. 16, No.1 (1985), pp.117-122.

Schniederjans, Marc J., "A Statistical Screening-Procedure for Goal Programming Algorithm Selection," *Socio-Economic Planning Sciences*, Vol. 20, No. 3 (1986), pp. 155-160.

Schniederjans, Marc J., Kwak, N. J. and Helmer, M. C., "An Application of Goal Programming to Resolve a Site Location Problem," *Interfaces*, Vol. 12, No. 3 (1983), pp. pp. 65-72.

Schniederjans, Marc J., Zorn, T. S. and Johnson, R. R., "Allocating Total Wealth: A Goal Programming Approach," *Computers and Operations Reserach*, Vol. 20, No. 7 (1993), pp. 679-685.

Schroeder, Roger G., "Resource Planning in University Management by Goal Programming," *Operations Reseach*, Vol. 22, No. 4 (1974), pp. 700-710.

Schuler, A. T. and Meadows, J. C., "Planning Resource Use on National Forests to Achieve Multiple Objectives," *Journal of Environmental Management*, Vol. 3 (1975), pp. 351-366.

Schuler, A. T., Webster, H. H. and Meadows, J. C., "Goal Programming in Forest Management," *Journal of Forestry*, Vol. 75, No. 6 (June 1977), pp. 320-324.

Schutt, Klaus P., "A Model to Support the Assessment of Subjective Probabilities," *Theory and Decision*, Vol. 12, No. 2 (1980), pp. 173-183.

Sealey, C. W. Jr., "Financial Planning with Multiple Objectives," *Financial Management*, Vol. 7, No. 4 (Winter 1978), pp. 17-23.

Sealey, C. W. Jr., "Commercial Bank Portfolio Management with Multiple Objectives," *Journal of Commercial Bank Lending*, Vol. 59, No. 6 (1977), pp. 39-48.

Seiford, L. and Yu, P. L., "Potential Solutions of Linear Systems - Multi-Criteria Multiple Constraint Levels Program," *Journal of Mathematical Analysis and Applications*, Vol. 69, No. 2 (1979), pp. 283-303.

Selen, Willem J. and Hott, David D., "A Mixed-Integer Goal-Programming Formulation of the Standard Flow-Shop Scheduling Problem," *Journal of the Operational Research Society*, Vol. 37, No. 12 (1986), pp. 1121-1128.

Selin, S. Z. and Rifai, A. K., "Constraint Partitioning and Variable Elimination in Goal Programming," *Industrial Management*, Vol. 24, No. 5 (1982), pp. 1-6.

Sen, P., Shi, W. B. and Caldwell, J. B., "Efficient Design of Panel Structures by a General Multiple Criteria Utility," *Engineering Optimization*, Vol. 14 (1989), pp. 287-310.

Sengupta, Jati K., "Stochastic Goal Programming with Estimated Parameters," *Zeitschrift fur Nationalokonomie-Journal of Economics*, Vol. 8, No. 1 (1981), pp. 58-65.

Sengupta, S., "Goal Programming Approach to a Type of Quality Control Problem," *Journal of the Operational Research Society*, Vol. 32, No. 3 (March 1981), pp. 207-211.

Sexton, T. R., Silkman, R. H. and Hogan, A. J., "Data Envelopment Analysis: Critique and Extensions," *New Directions for Program Evalution*, Vol. 32 (Winter 1986), pp. 73-105.

Shafer, S. M. and Rogers, D. F., "A Goal Programming Approach to the Cell Formulation Problem," *Journal of Operations Management*, Vol. 10, No. 1 (1991), pp. 28-43

Sharda, Ramesh, and Musser, Kathryn D., "Financial Future Hedging via Goal Programming," *Management Science*, Vol. 32, No. 8 (1986), pp. 933-947.

Sharif, M. N. and Agarwal, R. L., "Solving Multicriterion Integer Programming Problems," *Industrial Management*, Vol. 18, No. 1 (1976), pp. 17-23.

Sharma, J. K. and Sharma, M. M., "On Mixed Integer Solutions to Goal Programming Problems," *Indian Journal of Pure and Applied Mathematics*, Vol. 11, No. 3 (1980), pp. 314-320.

Shepherd, J. G., "Cautions on Non-linear Optimization: A New Technique for Allocation Problems," *Journal of the Operational Research Society*, Vol. 31 (1980), pp. 993-1000.

Shepherd, J. G., "Matching Fishing Capacity to the Catches Available: A Problem in Resource Allocation," *Journal of Agricultural Economic*, Vol. 32 (1981), pp. 331-340.

Sherali, H. D. and Soyster, A. L. "Preemptive and Non-preemptive Multi-objective Programming: Relationships and Counter Examples," *Journal of Optimization Theory and Applications*, Vol. 39 (1983), pp. 173-186.

Sherali, H. D., "Equivalent Weights for Lexicographic Multi-Objective Programs: Charaterizations and Computations," *European Journal of Operational Research*, Vol. 11, No. 4 (1982), pp. 367-379.

Sheshai, K. M., Harwood, G. B. and Harmanison, R. H., "Cost Volume Profit Analysis with Integer Goal Programming," *Management Accounting*, Vol. 59, No. 4 (1977), pp. 43-47.

Shih, C.-J. and Hagels, P., "Multicriteria Optimum Design of Belleville Spring Stack with Discrete and Integer Decision Variables," *Engineering Optimization*, Vol. 15, No. 1 (1989) pp. 43-56.

Shim, J.P. and Chun, S. G., "Goal Programming : The RPMS Network Approach," *Journal of the Operational Research Society*, Vol. 42, No. 1 (1991) pp. 83-93.

Shim, Jae K. and Siegel, J., "Quadratic Preferences and Goal Programming," *Decision Sciences*, Vol. 6, No. 4 (1975), pp. 662-669.

Shim, Jae K. and Siegel, J., "Sensitivity Analysis of Goal Programming with Pre-emption," *International Journal of Systems Science*, Vol. 11, No. 4 (April 1980), pp. 393-401.

Shim, Jae K., "A Survey of Quadratic Programmig Applications to Business and Economics," *International Journal of Systems Science*, Vol. 14, No. 1 (1983), pp. 105-115.

Shin, Wan S. and Ravindran, A., "Interactive Multiple Objective Optimization : Survey 1 - Continuous Case," *Computers and Operations Reserach*, Vol. 18, No. 1 (1991), pp. 97-114.

Siha, Samia, "A Decision Model for Selecting Mutually Exclusive Alternative Technologies," *Computers and Industrial Engineering*, Vol. 24, No. 3 (July 1993), pp. 459-475.

Simkin, Mark G., "Mathematical Programming for the State Earmarking Process," *Decision Sciences*, Vol. 8, No. 1 (1977), pp. 256-269.

Singh, N. and Agarwal, S. K., "Optimum Design of an Extended Octagonal Ring by Goal Programming," *International Journal of Production Research*, Vol. 21, No. 6 (1983), pp. 891-898.

Singh, N. and Kishore, N., "Prioritized Weighted Goal Programming Formulation of an Unbalanced Transportation Problem with Budgetary Constraints,"*Industrial Engineering Journal*, Vol. 20, No. 5 (May 1991), pp.16-22.

Singh, N. and Verma, A. P., "Optimization of Dressing Variables in a Single Point Diamond Dressing: A Goal Programming Approach," *Engineering Optimization*, Vol. 9 (1985), pp. 51-60.

Singh, N., Aneja, Y.P. and Rana, S.P., "Multiobjective Modelling and Analysis of Process Planning in a Manufacturing System," *International Journal of Systems Science*, Vol. 21, No. 4 (1990) pp. 621-630.

Singh, N., "Optimum Design of a Journal Bearing System with Mult-Objectives: A Goal Programming Approach," *Engineering Optimization*, Vol. 6 (1983), pp. 193-196.

Sinha, S. B. and Sastry, S. V. C., "A Goal Programming Model for Facility Location Planning," *Socio-Economic Planning Sciences*, Vol. 21, No. 4 (1987a), pp. 251-256.

Sinha, S. B. and Sastry, S. V. C., "General Mathematical Modelling of a Multi-Objective Community Stroage Facility System," *Socio-Economic Planning Sciences*, Vol 21, No. 1 (1987b), pp. 1-7.

Sinha, S. B., Rao, K. A. and Mangaraj, B. K., "Fuzzy Goal programming In Multi-Criteria Decision Systems: A Case Study in Agricultural Planning," *Socio-Economic Planning Sciences*, Vol. 22, No. 2 (1988), pp. 93-102.

Sinha, S. B., Sastry, S. V. C. and Misra, R. P., "Management of a Community Storage System: A Goal Programming Approach," *European Journal of Operational Research*, Vol. 34 (1988), pp. 92-98.

Siokas, D. and Vassiloglou, M., "A Goal Programming Model for Bank Assets and Liabilities Management," *European Journal of Operational Research*, Vol. 50, No. 1 (January 1991), pp. 48-60.

Slade, Stephen, "Generating Explanations for Goal-Based Decision Making," *Decision Sciences*, Vol. 23, No. 6 (1992), pp. 1440-1461.

Smith, L. D., "Planning Models for Budgeting Teaching Resources," *Omega*, Vol. 6, No. 1 (1978), pp. 83-88.

Smith, C. J., "Using Goal Programming to Determine Interest Group Disutility for Public Policy Choices," *Socio-Economic Planning Sciences*, Vol. 14, No. 3 (1980), pp. 117-120.

Solanki, R., "Generating the Noninferior Set in Mixed Integer Biobjective Linear Programs: An Application to a Location Problem," *Computers and Operations Research*, Vol. 18, No. 1 (1991), pp. 1-16.

Solomon Barry D. and Haynes, Kingsley E., "A Survey and Critique of Multiobjective Power Plant Citing Decision Rules," *Socio-Economic Planning Sciences*, Vol. 18, No. 2 (1984), pp. 71-79.

Soyibo, Adedoyin, "Goal Programming Methods and Applications: A Survey," *Journal of Information and Optimization Sciences*, Vol. 6, No. 3 (1985), pp. 247-264.

Soyibo, Adedoyin and Lee, Sang M., "A Multiobjective Planning Model for University Resource Allocation," *European Journal of Operational Research*, Vol. 27 (1986), pp. 168-178.

Soyster, A. L. and Lev, B., "An Interpretation of Fractional Objectives in Goal Programming as Related to Papers by Awerbuch, et al., and Hannan," *Management Science*, Vol. 24 (October 1978), pp. 1546-1549.

Specht, P. H., "Multicriteria Planning Model for Mental Health Services Delivery," *International Journal of Operations and Production Management*, Vol. 13, No. 9 (1993), pp. 62-71.

Spivey, W. A. and Tamura, H., "Goal Programming in Econometrics," *Navel Research Logistics Quaterly*, Vol. 17, No. 2 (1970), pp. 183-192.

Spronk, Jaap and Veeneklaas, Frank, "A Feasibility Study of Economic and Environmental Scenarios by Means of Interactive Multiple Goal Programming," *Regional Science and Urban Economics*, Vol. 13, No. 1 (1983), pp. 141-160.

Sridhar, U. and Raghavendra, B. G., "A Reference Level for Evaluation of Corporate Performance Using Goal Programming Models," *Policy and Information*, Vol. 12, No. 2 (1988), pp. 109-120.

Srinivasan, Venkat and Kim, Yong H., "Credit Granting: A Comparative Analysis of Classification Procedures," *Journal of Finance*, Vol 42, No. 3 (1987), pp. 665-683.

Srinivasan, Venkat and Shocker, Allan D., "Linear Programming Techniques for Multidimensional Analysis of Preferences," *Psychometrika*, Vol. 38, No. 3 (September 1973), pp. 337-369.

Srinivasan, Venkat and Thompson, G. L., "Determining Optimal Growth Paths in Logistics Operations," *Naval Research Logistics Quaterly*, Vol. 19 (1972), pp. 575-599.

Stadje, W., "On the Relationship of Goal Programming and Utility Functions," *Zeitschrift fur Operations Research*, Serie A & B, Vol. 23, No. 1 (1979), pp. A61-A69.

Stam, A., "Extensions of Mathematical Programming-Based Classification Rules : A Multicriteria Approach," *European Journal of Operational Research*, Vol. 48, No. 3 (1990) pp. 351-361.

Stancu-Minasian, I. M. and Tigan, S., "A Stochastic Approach to Some Linear Fractional Goal Programming Problems," *Kybernetika*, Vol. 24, No. 2 (1988), pp. 139-149.

Stern, H. I., "A Goal Programming Approach to Planning Population Balance in a Multiregional System," *Environment and Planning Analysis*, Vol. 6, No. 4 (1974), pp. 431-437.

Steuer, Ralph E., "Goal Programming Sensitivity Analysis Using Interval Penalty Weights," *Mathematical Programming*, Vol. 17, No. 1 (1979), pp. 16-31.

Steuer, Ralph E., "Multiple Criterion Function Goal Programming Applied to Managerial Compensation Planning," *Computers and Operations Reserach*, Vol. 10, No. 4 (1983), pp. 299-309.

Stewart, T. J., "A Multi-Criteria Decision Support System for R&D Project Selection," *Journal of the Operational Research Society*, Vol. 42, No. 1 (1991) pp. 17-26.

Stewart, T. J., "A Descriptive Approach to Multi-Criteria Decision Making," *Journal of the Operational Research Society*, Vol. 32, No. 1 (1981) pp. 45-53.

Stewart, T. J. and Ittman, "Two-Stage Optimization in a Transportation Problem," *Journal of the Operational Research Society*, Vol. 30 (1979), pp. 897-904.

Stewart, T. J., "Interactive Utility Assessment in Multicriteria Decision Analysis Using Implicit Trade-Off Information," *Journal of the Operational Research Society*, Vol. 39, No. 3 (1988), pp. 285-298.

Stone, B. K. and Reback, R., "Constructing a Model for Managing Portfolio Revisions," *Journal of Bank Research*, Vol. 6, No. 1 (1975), pp. 48-60.

Sueyoshi, Toshiyuki, "Estimation of Stochastic Frontier Cost Function Using Data Envelopment Analysis: An Application to the AT&T Divestiture," *Journal of the Operational Research Society*, Vol. 42, No. 6 (June 1991), pp. 463-477.

Sueyoshi, Toshiyuki, "Goal Programming Approach for Regression Median," *Decision Sciences*, Vol. 20, No. 4 (Fall 1989), pp. 700-712.

Sullivan, R. S. and Fitzsimmons, J. A., "A Goal Programming Model for Readiness and the Optimal Redeployment of Resources," *Socio-Economic Planning Sciences*, Vol. 12, No. 5 (1978), pp. 215-220.

Sundaram, R. M., "An Application of Goal Programming Technique in Metal Cutting," *International Journal of Production Research*, Vol. 16, No. 5 (1978), pp. 375-382.

Sushil Dewan, S. and Vrat, P., "Waste Management Policy Analysis and Growth Monitoring: An Integrated Approach to Perspective Planning," *International Journal of Systems Science*, Vol. 20, No. 6 (June 1989), pp. 907-926.

Sushil, Dewan, S. and Agrawal, V. K., "Application of Goal Programming for Capacity Waste Minimization," *International Journal of Operations and Production Management*, Vol. 9, No. 3 (1989), pp. 26-34.

Sushil Dewan, S., "Application of Physical System Theory and Goal Programming to Modelling and Analysis of Waste Management in National Planning," *International Journal of Systems Science*, Vol. 24, No. 5 (1993), pp. 957-984.

Sutcliffe, C. M. S., and Board, J. L. G., "Designing Secondary-School Catchment Areas Using Goal Programming," *Environment and Planning Analysis*, Vol. 18, No. 5 (1986), pp. 661-675.

Sutcliffe, Charles, Board, John, and Cheshire, Paul, "Goal Programming and Allocating Children to Secondary Schools in Reading," *Journal of the Operational Research Society*, Vol. 35, No. 8 (1984), pp. 719-730.

Sutcliffe, Charles, Board, John, and Cheshire, Paul, "Naive Weighting in Non-preemptive Goal Programming: A Reply," *Journal of the Operational Research Society*, Vol. 36 (1985), pp. 648-649.

Suzuki, S. and Matsuda, S., "Structure/Control Design Synthesis of Active Flutter Suppression System by Goal Programming," *Journal of Guidance, Control, and Dynamics*, Vol. 14, No. 6 (1991), pp. 1260-1268.

Suzuki, S. and Yoshizawa, T., "Multiobjective Trajectory Optimization by Goal Programming with Fuzzy Decisions," *Journal of Guidance, Control, and Dynamics*, Vol. 17, No. 2 (March-April 1994), pp. 297-303.

Svestka, Joseph A., "MOCRAFT : A Professional Quality Micro-computer Implementation of CRAFT with Multiple Objectives," *Computers and Industrial Engineering*, Vol. 18, No. 1 (1990) pp. 13-22.

Sweeney, D. J., Winkofsky, E. P., Roy, P. and Baker, N. R., "Composition vs. Decomposition: Two Approaches to Modeling Organizational Decision Processes," *Management Science*, Vol. 24 (1978), pp. 1491-1499.

Szidarovsky, Ference, and Duckstein, Lucien, "Dynamic Multiobjective Optimization: A Framework with Application to Regional Water and Mining Management," *European Journal of Operational Research*, Vol. 24, No. 2 (1986), pp. 305-317.

Tabucanon, M. T. and Mukyangkoon, S., "Multi-objective Microcomputer-Based Interactive Production Planning," *International Journal of Production Research*, Vol. 23 (1985), pp. 1001-1023.

Taner, Orhan V. and Koksalan, M. Murat, "Experiments and an Improved Method for Solving the Discrete Alternative Multiple-Criteria Problem," *Journal of the Operational Research Society*, Vol. 42, No. 5 (1991) pp. 401-411.

Tanner, L., "Selecting a Text-Processing System as a Qualitative Multiple Criteria Problem," *European Journal of Operational Research*, Vol. 50, No. 2 (1991) pp. 179-187.

Tayi, Giri K., "A Polonomial Goal Programming Approach to a Class of Quality Control Problems," *Journal of Operations Management*, Vol. 5, No. 2 (1985), pp. 237-246.

Tayi, Giri K. and Gangolly, Jagdish, "Integration of Auditor Preferences and Sampling Objectives: A Polynomial Goal-Programming Perspective," *Journal of the Operational Research Society*, Vol. 36, No. 10 (1985), pp. 951-957.

Tayi, Giri K. and Leonard, Paul A., "Bank Balance-sheet Management: An Alternative Multi-objective Model," *Journal of the Operational Research Society*, Vol. 39, No. 4 (1988), pp. 401-410.

Taylor, Bernard W. III and Anderson, P. F., "Goal Programming Approach to Marketing/Production Planning," *Industry Marketing Management*, Vol. 8, No. 2 (1979), pp. 136-144.

Taylor, Bernard W. III and Keown, A. J., "A Goal Programming Application of Capital Project Selection in the Production Area," *AIIE Transactions*, Vol. 10, No. 1 (1978a), pp. 52-57.

Taylor, Bernard W. III and Keown, A. J., "A Mixed-Integer Goal Programming Model for Capital Budgeting within a Police Department," *Computers, Environment and Urban Systems*, Vol. 6, No. 3 (1981), pp. 171-181.

Taylor, Bernard W. III and Keown, A. J., "Planning Urban Recreational Facilities with Integer Goal Programming," *Journal of the Operational Research Society*, Vol. 29, No. 8 (1978b), pp. 751-758.

Taylor, Bernard W. III, Davis, K. R. and North, R. M., "Approaches to Multiobjective Planning in Water Resources Projects," *Water Resources Bulletin*, Vol. 11, No. 5 (1975), pp. 999-1008.

Taylor, Bernard W. III, Davis, K. R. and Ryan, L. J., "Compliance with the Occupational Safety and Health Act: A Mathematical Framework," *Decision Sciences*, Vol. 8, No. 5 (1977), pp. 677-691.

Taylor, Bernard W. III, Keown, Arthur J., and Greenwood, Allen G., "An Integer Goal Programming Model for Determining Military Aircraft Expenditures," *Journal of the Operational Research Society*, Vol. 34, No. 5 (1983), pp. 379-390.

Taylor, Bernard W. III, Moore, Laurence J. and Clayton, Edward R., "R and D Project Selection and Manpower Allocation with Integer Nonlinear Goal Programming," *Management Science*, Vol. 28, No. 10 (1982), pp. 1149-1158.

Taylor, Bernard W. III, Moore, Laurence J., Clayton, Edward R., Davis, K. Roscoe and Rakes, Terry R., "An Integer Nonlinear Goal programming Model for the Development of State Highway Patrol Units," *Management Science*, Vol. 31, No. 11 (1985), pp. 1335-1347.

Taylor, Bernard W. III, "Urban Recreational Planning: A Proposed Methodology," *Omega*, Vol. 5, No. 5 (1977), pp. 618-619.

Teghem J., Dugrane, D., Thauvoye, M. and Kunsch, P., "STRANGE: An Interactive Method for Multi-Objective Linear Programming Under Uncertainty," *European Journal of Operational Research*, Vol. 26 (1986), pp. 65-82.

Tersine, R. J., "Organizational Objectives and Goal Programming: A Convergence," *Managerial Planning*, Vol. 25, No. 2 (1976), pp. 27-32.

Thanassoulis, E., "Selecting a Suitable Solution Method for a Multi Objective Programming Capital Budgeting Problem," *Journal of Business Finance and Accounting*, Vol. 12, No. 3 (1985), pp. 453-471.

Thomas, R. W., "Developments in Mathematical Programming Models and Their Impact on th Spatial Allocation of Education Resources," *Progress-in-Human-Geography*, Vol. 11, No. 2 (June 1987), pp. 207-226.

Thore, S., Nagurney, A. and Pan, J., "Generalized Goal Programming and Variational Inequalities," *Operations Reseach* Letters, Vol. 12, No. 4 (1992), pp. 217-226.

Tingley, Kim M. and Liebman, Judith S., "A Goal Programming Example in Public Health Resource Allocation," *Management Science*, Vol. 30, No. 3 (1984a), pp. 279-289.

Tiwari, R. N., Dharmar, S. and Rao, J. R., "Priority Structure in Fuzzy Goal Programming," *Fuzzy Sets and Systems*, Vol. 19, No. 3 (July 1986), pp. 251-260.

Tiwari, R. N., Dharmar, S. and Rao, J. R., "Fuzzy Goal Programming-An Additive Model," *Fuzzy Sets and Systems*, Vol. 24, No. 1 (1987), pp. 27-34.

Trivedi, V. M., "A Mixed-Integer Goal Programming Model for Nursing Service Budgeting," *Operations Reseach*, Vol. 29, No. 5 (1981), pp. 1019-1034.

Tseng, C. H. and Lu, T. W., "Minimax Multiobjective Optimization in Structural Design," *International Journal of Numerical Methods in Engineering*, Vol. 30, No. 6 (1990), pp. 1213-1228.

Turshen, I. J. and Wester, K. W., "Allocation of Funds to Local Transportation Activities: Goal Programming and the Virginia Rideshare Program," *Transportation Journal*, Vol. 26, No. 6 (Winter 1986), pp. 61-70.

Tyagi, M. S. and Swarup, K., "Rural Economic Development and Employment Potential: A Mathematical Programming Approach," *Productivity*, Vol. 20, No. 1 (1979), pp. 63-75.

Utar, K. and Schoenfled, H. M., "Integrating Production Scheduling, Capacity Acquisition, and/or Abandonment and Make-Buy Decisions," *International Management Review*, Vol. 12, No. 3 (1973), pp. 99-115.

Van Crombrugge, M. and Thompson, W., "Optimization of the Transmitting Characteristics of a Tonpilz-type Transducer by Proper Choice of Impedance Matching Layers," *Journal of the Acoustical Society of America*, Vol. 77 (1985), pp. 747-752.

Van Hulle, M. M., "A Goal Programming Network for Linear Programming," *Biological Cybernetics*, Vol. 65, No. 4 (1991a), pp. 243-252.

Van Hulle, M. M., "A Goal Programming Network for Mixed Integer Linear Programming: A Case Study for the Job-Shop Scheduling Problem," *International Journal of Neural Systems*, Vol. 2, No. 3 (1991b), pp. 201-209.

Varshney, K. G. and Rao, Y. P., "Multiobjective Crop Planning Using Goal Programming- A Case Study," *Journal of the Institute of Engineers* (India), Part AG: Agricultural Engineering Division, Vol. 69, Part 2 (January 1989), pp. 58-65.

Venugopal, S. and Mohanty, R. P., "A Multigoal Analytic Formulation for Plant Layout Problem," *Journal of the Institution of Engineers*, Vol. 63 (1982), pp. 48-56.

Vickery, Shawnee K. and Markland, Robert E., "Integer Goal Programming for Multistage Lot Sizing: Experimentation and Implementation," *Journal of Operations Management*, Vol. 5, No. 2 (1985), pp. 169-182.

Vickery, Shawnee K. and Markland, Robert E., "Multi-Stage Lot Sizing in a Serial Production System," *International Journal of Production Research*, Vol. 24, No. 3 (1986), pp. 517-534.

Vinso, J. D., "Financial Planning for the Multinational Corporation with Multiple Goals," *Journal of International Business Studies*, Vol. 13, No. 3 (Winter 1982), pp. 43-58.

Vrat, Prem and Kriengkrairut, Charoen, "A Goal Programming Model for Project Crashing with Piecewise Linear Time-Cost Trade-Off," *Engineering Costs and Production Economics*, Vol. 10, No. 2 (1986), pp. 161-172.

Wacht, R. F. and Whitford, D. T., "A Goal Programming Model for Capital Investment Analysis in Nonprofit Hospitals," *Financial Management*, Vol. 5, No. 2 (Summer 1976), pp. 37-46.

Walker, H. D., "An Alternative Approach to Goal Programming," Canadian Journal of Forest Research, Vol. 15, No. 2 (1985), pp. 319-325.

Walker, J., "An Interactive Method as an Aid in Solving Multi-Objective Mathematical Programming Problems," *European Journal of Operational Research*, Vol. 2, No. 5 (1978), pp. 341-349.

Walker, M. C. and Chandler, G. G., "Equitable Allocation of Credit Union Net Revenues - Goal Programming Approach," *Journal of Economics and Business*, Vol. 31, No. 1 (1978), pp. 63-69.

Walleniius, H., "Optimizing Macroeconomic Policy: A Review of Approaches and Applications," *European Journal of Operational Research*, Vol. 10, No. 3 (1982), pp. 221-228.

Walters, A., Mangold, J. and Haran, E. G. P., "A Comprehensive Planning Model for Long-Range Academic Strategies," *Management Science*, Vol. 22, No. 7 (1976), pp. 727-738.

Wang, Y. M., "Nonlinear Goal Programming Model and Algorithm of Maximum Entropy Image Reconstruction From Projections," *Journal of Hunan University*, Vol. 13, No. 2 (1986), pp. 12-19.

Wascher, G., "An LP-based Approach to Cutting Stock Problems with Multiple Objectives," *European Journal of Operational Research*, Vol. 44, No. 2 (1990) pp. 175-184.

Weigel, H. S. and Wilcox, S. P., "The Army's Personnel Decision Support System," *Decision Support Systems*, Vol. 9, No. 3 (April 1993), pp. 281-306.

Weistroffer, H. R., "An Interactive Goal Programming Method for Non-Linear Multiple-Criteria Decision-Making Problems," *Computers and Operations Reserach*, Vol. 10, No. 4 (1983), pp. 311-320.

Weistroffer, H. R., "Multiple Criteria Decision Making with Interactive Over-Achievement Programming," *Operations Research Letters*, Vol. 1 (1982), pp. 241-245.

Weithman, A. S. and Ebert, R. J., "Goal Programming to Assist in Decision Making," *Fisheries*, Vol. 6, No. 1 (January-February 1981), pp. 5-8.

Welling, P., "A Goal Programming Model for Human Resources Accounting in a CPA Firm," *Accounting, Organizations and Society*, Vol. 2, No. 4 (1977), pp. 307-316.

Welman, Ulf P., "Comments on Goal Programming for Aggregate Planning," *Management Science*, Vol. 22, No. 6 (1976), pp. 708-712.

Werczberger, E., "A Goal Programming Model for Industrial Location Involving Environmental Considerations," *Environment and Planning Analysis*, Vol. 8, No. 2 (1976), pp. 173-188.

Werczberger, E., "Multi-Objective Linear Programming with Partial Ranking of Objectives," *Socio-Economic Planning Sciences*, Vol. 15 (1981), pp. 331-340.

Werczberger, E., "Planning in an Uncertain Environment: Stochastic Goal Programming Using the Versatility Criterion," *Socio-Economic Planning Sciences*, Vol. 18, No. 6 (1984), pp. 391-398.

Wheeler, B. M. and Russell, J. R. M., "Goal Programming and Agricultural Planning," *Operational Research Quaterly*, Vol. 28 (1977), pp. 21-32.

White, D.J., "A Bibliography on the Applications of Mathematical Programming Multiple-Objective Methods," *Journal of the Operational Research Society*, Vol. 41, No. 8 (1990), pp. 669-692.

White, D.J., "A Characterisation of the Feasible Set of Objective Functon Vectors in Linear Multiple Objective Problem," *European Journal of Operational Research*, Vol. 52, No. 3 (1991) pp. 361-366.

Whitford, D. T. and Davis, W. J., "A Generalized Hierarchical Model of Resource Allocation," *Omega*, Vol. 11 (1983), pp. 279-291.

Widhelm, W. B., "Extensions of Goal Programming Models," *Omega*, Vol. 9, No. 2 (1981), pp. 212-214.

Wierzbicki, A. P., "A Mathematical Basis for Satisficing Decision Making," *Mathematical Modelling*, Vol. 3, No. 5 (1982), pp. 391-405.

Wijngaard, J., "Aggregation in Manpower Planning," *Management Science*, Vol.29, No.12 (1983), pp.1427-1435.

Williams, A. J., "Causes of Purchasing Myopia," *Industrial Marketing and Purchasing*," Vol. 2, No. 1 (1987), pp. 26-29.

Williams, Roy H. and Zigli, Ronald M., "Ambiguity Impedes Quality in the Service Industries," *Quality Progress*, Vol. 20, No. 7 (1987), pp. 14-17.

Willis, C. E. and Perlack, R. D., "A Comparison of Generating Techniques and Goal Programming for Public Investment, Multiple Objective Decision Making," *American Journal of Agricultural Economics*, Vol. 62, No. 1 (1980a), pp. 66-74.

Willis, C. E. and Perlack, R. D., "Multiple Objective Decision Making: Generating Techniques or Goal Programming," *Journal of Northeast Agricultural Economic Council*, Vol. 9, No. 1 (1980b), pp. 1-6.

Wilson, G. R. and Jain, H. K., "An Approach to Postoptimality and Sensitivity Analysis of Zero-One Goal Programs," *Naval Research Logistics*, Vol. 35, No. 1 (1988), pp. 73-84.

Wilson, F. R. and Gonzalez, H., "Mathematical Programming for Highway Project Analysis," *Journal of Transportation Engineering*, Vol. 11 (1985), pp. 162-171.

Wilson, J. M., "Alternative Formulations of a Flow-Shop Scheduling Problem," *Journal of the Operational Research Society*, Vol. 40, No. 4 (April 1989), pp. 395-399.

Wilson, J. M., "The Handling of Goals in Marketing Problems," *Management Decision*, Vol. 13 (1975), pp. 175-180.

Wilson, Richard M. and Gibberd, Robert W., "Combining Multiple Criteria for Regional Resource Allocation in Health Care Systems," *Mathematical Computer Modeling*, Vol. 13, No. 8 (1990) pp. 15-27.

Wilstead, W. D., Hendrick, T. E. and Stewart, T. R., "Judgment Policy Capturing for Bank Loan Decisions: An Approach to Developing Objective Functions for Goal Programming Models," *Journal of Management Studies*, Vol. 12, No. 2 (1975), pp. 210-225.

Wolamowsky, Yonah, Epstein, S. and Dickman, D., "Optimization in Multiple-Objective Linear Programming Problems with Pre-Emptive Priorities," *Journal of the Operational Research Society*, Vol. 41, No. 4 (1990), pp. 351-356.

Wright, J., Revelle, C. and Cohon, J., "A Multiobjective Integer Programming Model for the Land Acquistion Problem," *Regional Science and Urban Economics*, Vol. 13, No. 1 (1983), pp. 31-53.

Yang, Y.B., Chen, C. and Zhang, Z. J., "The Interactive Step Trade-off Method (ISTM) for Multiobjective Optimization," *IEEE Transations on Systems, Man, and Cybernetics*, Vol. 20, No. 3 (1990) pp. 688-695.

Yazadanian, A. and Peralta, R. C., "Maintaining Target Groundwater Levels Using Goal-Programming: Linear and Quadratic Methods," *Transactions on America Society of Agricultural Engineering*, Vol. 29, No. 4 (1986), pp. 995-1004.

Zahedi, Fatemeh, "Qualitative Programming For Selection Decisions," *Computers and Operations Reserach*, Vol. 14, No. 5 (1987), pp. 395-407.

Zaloom, V., Tolga, A. and Chu, H., "Bank Funds Management by Goal Programming," *Computers and Industrial Engineering*, Vol. 11, Nos. 1-4 (1986), pp. 132-135.

Zanakis, Stelios H. and Lawrence, K. D., "Effective Shift Allocation in Multi-product Lines: A Mixed Integer Programming Approach," *Operational Research Quaterly*, Vol. 28 (1977), pp. 1013-1021.

Zanakis, Stelios H. and Maret, M. W., "A Markovian Goal Programming Approach to Aggregate Manpower Planning," *Journal of the Operational Research Society*, Vol. 32, No. 1 (January 1981a), pp. 55-63.

Zanakis, Stelios H. and Maret, M. W., "A Note on Aggregate Manpower Planning Using A Markovian Goal Programming Approach," *Journal of the Operational Research Society*, Vol. 32, No. 12 (December 1981b), pp. 1143-1156.

Zanakis, Stelios H. and Smith, J. S., "Chemical Production Planning via Goal Programming," *International Journal of Production Research*, Vol. 18, No. 6 (1980), pp. 687-697.

Zanakis, Stelios H. and Gupta, Sushil K., "A Categorized Bibliographic Survey of Goal Programming," *Omega*, Vol. 13, No. 3 (1985), pp. 211-222.

Zanakis, Stelios H., "A Staff to Job Assignment (Partitioning) Problem with Multiple Objectives," *Computers and Operations Reserach*, Vol. 10, No. 4 (1983), pp. 357-374.

Zanakis, Stelios H., Mavrides, Lazaros P., and Roussakis, Emmanuel N., "Applications of Management Science in Banking," *Decision Sciences*, Vol. 17, No.1 (1986), pp. 114-128.

Zanakis, Stelios H., "Social Program Evaluation and Fund Allocation: A 0-1 Goal Programming Case Study," *Management Science* and Policy Analysis, Vol. 5, No. 1 (1987), pp. 40-51.

Zeleny, Milan, "The Pros and Cons of Goal Programming," *Computers and Operations Reserach*, Vol. 8, No. 4 (1981), pp. 357-359.

Zhu, Z. and McKnew, M. A., "A Goal Programming Workload Balancing Optimization Model for Ambulance Allocation: An Application to Shanghai, P. R. China," *Socio-Economic Planning Sciences*, Vol. 27, No. 2 (1993), pp. 137-148.

Zimmermann, H. J., "Fuzzy Mathematical Programming," *Computers and Operations Reserach*, Vol. 10, No. 4 (1983), pp. 291-298.

Zimmermann, H. J., "Fuzzy Programming and Linear Programming with Several Objectives," *Fuzzy Sets and Systems*, Vol. 1 (1978), pp. 45-55.

Zionts, S. and Wallenius, J., "An Interactive Programming Method for Solving the Multiple Criteria Problem," *Management Science*, Vol. 22, No. 6 (1976), pp. 652-663.

INDEX